Catalog of
Cosmic Gamma-Ray Bursts
from the
KONUS Experiment Data

by

E. P. MAZETS *et al*.

Reprinted from

Astrophysics and Space Science, Vol. 80, No. 1

D. Reidel Publishing Company

P.O. Box 17, 3300 AA Dordrecht, Holland

ISBN-13: 978-90-277-9050-7   e-ISBN-13: 978-94-009-7960-4
DOI: 10.1007/ 978-94-009-7960-4

TABLE OF CONTENTS

CATALOG OF COSMIC GAMMA-RAY BURSTS FROM THE KONUS EXPERIMENT DATA

Preface

Shortly after the first announcement of the discovery of gamma-ray bursts in 1973, a conference was held at the Los Alamos National Laboratories in New Mexico to discuss the current results and plan future experiments. It was clear then that one of the priorities for future research in the field was the precise localisation of these events, since this would allow careful searches to be made in the optical, soft X-ray, and radio ranges. The years that immediately followed saw the publication of a large number of theoretical and experimental papers on gamma-ray bursts, the latter based on experiments built for more traditional X- and gamma-ray astronomy observations. Although the number of publications eventually waned, several small groups of experimenters in France, the United States, and the Soviet Union continued to work on the construction of dedicated gamma-ray burst detectors. Their goal had been outlined at the Los Alamos meeting: to create an interplanetary network of gamma-ray burst detectors. The first spacecraft in this network was Helios-B, which carried an instrument into heliocentric orbit in 1976. In 1978, Pioneer Venus Orbiter established a point in the network in a Venus orbit, while Prognoz 7, Venera 11, and Venera 12 carried a total of 9 instruments into Earth and heliocentric orbits. ISEE-3 provided observations from a Lagrange point halo orbit.

The result was an astonishing harvest of data on gamma-ray bursts. While a team of American, French, and Soviet experimenters pooled their data to establish arc minute localization of intense events such as those of 19 November 1978, 5 March 1979, and 6 April 1979, Mazets and his co-workers in Leningrad began independently to establish, with a large collection of data on both weak and intense events, both the log N-log S relation, and the spatial distribution of bursts.

By 1979, considerable progress had been made in the study of these transient events, and it seemed appropriate to bring together once more the scientists working in this field, many of whom, collaborating by letter and telex, had never had the chance to meet one another. A meeting sponsored by the CNES and held in November 1979 at the Centre d'Etude Spatiale des Rayonnements in Toulouse, France, drew almost all of the workers in the field together for four days of presentations and workshop comparisons of data. It was a special pleasure to have the participation of Dr Mazets who, through the use of his KONUS experiments, was able to advance our understanding of gamma-ray bursts so quickly.

It has now been almost two years since that meeting, and exciting discoveries have been made almost continuously during that period. The debate over the origin of the 5 March 1979 event, the discovery of radio and X-ray sources in the field of the 19 November 1978 gamma-ray burst, the lack of optical counterparts down to 22nd magnitude for the 6 April 1979 event, the discovery of intense features in the energy spectra of gamma-ray bursts – all demonstrate that the study of gamma-ray bursts

Astrophysics and Space Science **80** (1981) 1–2. 0004–640X/81/0801–0001$00.30.

continues to be extremely rewarding. Experiments on future missions, such as the two 1981 Veneras, Gamma Ray Observatory, and the International Solar Polar Mission, will undoubtedly provide new and unexpected discoveries, but the work of the Leningrad group is likely to remain an important contribution to the field for some time to come.

C.E.S.R.   KEVIN HURLEY
Toulouse, France
June 9, 1981

CATALOG OF COSMIC GAMMA-RAY BURSTS FROM THE KONUS EXPERIMENT DATA

Parts I and II

E. P. MAZETS, S. V. GOLENETSKII, V. N. IL'INSKII, V. N. PANOV,
R. L. APTEKAR, YU. A. GUR'YAN, M. P. PROSKURA, I. A. SOKOLOV,
Z. YA. SOKOLOVA, and T. V. KHARITONOVA

A. F. Ioffe Physical-Technical Institute, U.S.S.R. Academy of Sciences, Leningrad, U.S.S.R.

and

A. V. DYATCHKOV and N. G. KHAVENSON

Institute of Cosmic Research, U.S.S.R. Academy of Sciences, Moscow, U.S.S.R.

(Received 2 July, 1980)

Abstract. Data are presented on the temporal structure, fluxes, energy spectra and coordinates of the sources of gamma-ray bursts detected in the KONUS experiment on Venera 11 and Venera 12 space probes in the period September 1978 to May 1979. The statistical distributions of gamma bursts in duration, intensity, and peak power, as well as the distribution of the burst sources over the celestial sphere presented are based on the updated KONUS information obtained until February 1980.

1. Introduction

The recently published paper by Mazets and Golenetskii (1981) contains a review of the major results obtained in the KONUS experiment. The present paper contains a sufficiently complete information on the gamma bursts detected in the experiment in the period September 1978 to May 1979 which had been given in Parts I and II of our Catalog (Mazets *et al.*, 1979a, 1979b). The statistical distributions of gamma bursts in duration T_b, intensity S ($E_\gamma > 30$ keV) and peak power $P(E_\gamma > 30$ keV) and the distributions of the burst sources over the celestial sphere have been constructed from a larger data base obtained until February 1980. Information pertaining to the gamma bursts detected in the last part of this period is scheduled to be published in the subsequent parts of the Catalog. The authors hope that the information presented will turn out to be useful in the comparision and interpretation of the results of gamma-burst studies both already done and being carried out on other spacecraft.

2. Instrumentation

The principal characteristics of the KONUS instrument and the conditions of observations have been described by Mazets *et al.* (1979c) and Mazets and Golenetskii (1981). We present here only the basic performance features needed for a proper understanding of the data in the Catalog.

The KONUS instrument operates in the triggered mode involving periodic measurement of the intensity and spectrum of cosmic and instrumental background in each of the six scintillation counters making up the sensor system. With the gamma flux exceeding the background by 6σ, a gamma-burst detection circuit is triggered, after which a time analyzer records the time profile of the burst in the energy range 50–150 keV during 2 s with a resolution 15.625 ms, then during 32 s with a 0.25 s resolution, and during the next 32 s with a 1 s resolution. The 8 s prehistory of the burst is also recorded with a 0.25 s resolution. A pulse-height analyzer records eight energy spectra in the range 30 keV – 2 MeV during eight consecutive time intervals of 4 s each.

An essential feature of the instrument is its capability of determining the direction of arrival of the gamma burst from one spacecraft by means of a system of six gamma-ray detectors possessing anisotropic angular sensitivity. Unambiguous localization of a source within a small area on the celestial sphere of a few degrees on a side is possible only from a spacecraft stabilized around the three axes in space. In observations carried out from a spacecraft with the direction of only one of its axes stabilized, the source field spreads into a ring on the celestial sphere described about the direction of the stabilized axis. Combining information obtained by this technique and by ordinary triangulation, where the direction to the source is derived from the difference in the gamma-burst arrival times measured on two spacecraft, permits one to obtain unambiguous, double-valued or, in the worst case, annular localization of the gamma-burst source, depending on the actual conditions of observation, event intensity, and the operating mode of the instrument and the spacecraft involved (Mazets *et al.*, 1979d).

3. Observations

85 gamma bursts have been detected in the experiment altogether in the 217 days of observation from September 1978 to May 1979. Forty-nine bursts have been detected in 107 days of simultaneous operation of the instrumentation on both spacecraft. Out of this number, seven weak bursts close in intensity to the detection threshold have been revealed on one spacecraft only.

143 gamma bursts have been recorded during the total observational period in the KONUS experiment in 384 days until February 1980.

The list and the characteristics of 85 gamma bursts are presented in Table I.

The first column of Table I contains dates of gamma-burst occurrence taken in what follows as a notation of the event. The times T_0 at which the gamma-burst detection circuit on Venera 11 and Venera 12 were triggered, are given in columns 2 and 3. Column 4 gives the burst intensity S which is the energy flux in the range $E\gamma > 30$ keV integrated over time of event. These figures were obtained from the measured spectra. In cases where a burst lasted for more than 32 s, the contribution from the final stage of the burst was derived from the burst profile recorded in the 50–150 keV region with corrections for the total region $E\gamma > 30$ keV calculated under the assumption of no further evolution of the spectrum.

Unambiguous localization of 30 sources responsible for 35 bursts detected in the period until May 1979, has been achieved; double-valued, or annular, localization has been obtained for a number of other bursts. Columns 5 and 6 of Table I contain the characteristics of the triangulation ring obtained from the burst arrival time difference for the two spacecraft (here and in what follows, the equatorial coordinates are given for the epoch 1950.0). For unambiguously localized sources, columns 7 and 8 give the mean source location in the equatorial and galactic coordinates; the coordinates of the quadrangular localization box are presented in column 9.

4. Time Profiles

Figures 1 to 87 display the time profiles of the gamma bursts detected. The graphs show the count rate in the energy range 50–150 keV vs. time with a 0.25 s resolution. The time T is reckoned from the time T_0 when the burst detection circuit was triggered. Points at $T - T_0 < 0$ represent the prehistory of the gamma burst recorded with the same detector. After $T - T_0 = 34$ s the time intervals in the histogram become four times longer, indicating the transfer of the time analyzer to the 1 s resolution operating mode. The horizontal dashed line shows the mean background level measured before and after a gamma burst. In cases of sufficient statistical reliability, the initial burst stage is shown with an 1/64 s resolution. For short bursts, only 1/64 resolution time profiles are presented.

5. Energy Spectra

The differential energy spectra of gamma bursts are presented in Figures 88 to 162. For some small bursts close in intensity to the detection threshold, the corresponding measurement errors are too large to permit presentation of the spectral data here. As seen from the graphs, the burst gamma-ray spectra may differ markedly; however, there are events with practically identical spectra. Some bursts reveal pronounced evolution of the spectrum in time. For such bursts we present some of the spectra obtained in several subsequent 4 s time intervals after T_0. The numbers of these intervals are specified near the corresponding curves in the graphs.

The most remarkable illustration of evolution of the energy spectrum during the burst is offered by the 19 November 1978 event (Figure 115). The general pattern of evolution, seen when going from the initial to the final burst stages (Figure 163), consists apparently in a substantial softening of the spectra.

6. Statistical Distributions

As already mentioned, the statistical distributions presented here are based on all the KONUS data obtained until February 1980.

TABLE I

Gamma-Burst Observations in the KONUS Experiment, September 1978–May 1979

Date of event	T_0, h.m.s. UT		$S(E\gamma > 30$ keV$)$, erg cm^{-2}	Triangulation ring		Mean source position		Box coordinates			
	V-11	V-12		Ring center α δ	Ring radius	α δ	l'' b''	α δ			
(1)	(2)	(3)	(4)	(5)	(6)	(7)	(8)	(9)			
14.09.78a	11.42.46	No data	9.2×10^{-6}			249.9 −3.1	13.1 26.7	252.0 −5.5	247.3 −4.6	252.3 −1.5	247.5 −0.6
14.09.78b	16.42.12	No data	1.7×10^{-5}								
16.09.78	No data	10.03.26	4.6×10^{-6}								
18.09.78	19.49.11	No data	2.8×10^{-5}								
21.09.78a	03.55.55	No data	1.8×10^{-5}								
21.09.78b	20.22.04	No data	1.0×10^{-6}								
22.09.78	12.34.06	No data	2.1×10^{-6}								
30.09.78	06.47.24	06.47.21	1.8×10^{-5}	232.344 36.459	61.9 ± 3.0						
4.10.78	00.47.19	Not detected	6.0×10^{-7}								
5.10.78	00.50.25	00.50.20	6.3×10^{-6}	56.402 −42.436	52.2 ± 2.0						
6.10.78a	10.59.50	10.59.52	2.6×10^{-5}	237.587 44.070	75.4 ± 2.5						
6.10.78b	14.25.19	14.25.17	1.7×10^{-5}	57.709 −44.229	76.0 ± 1.3						
11.10.78	10.52.33	Not detected	1.3×10^{-6}								
12.10.78a	01.06.20	01.06.21	9.3×10^{-6}	242.169 49.815	71.0 ± 1.3						
12.10.78b	17.13.49	17.13.47	4.0×10^{-6}	62.702 −50.425	72.0 ± 1.3						
13.10.78	14.54.41	No data	1.1×10^{-6}								
22.10.78	23.24.32	No data	2.0×10^{-6}								
23.10.78	17.23.35	No data	6.7×10^{-6}								

Date											
25.10.78	23.54.00	23.53.56	4.0×10^{-6}	73.051 / −60.026	49.5 ± 1.4	41.0 / −16.0	195.3 / −61.0	46.0 / −21.0	46.0 / −11.0	36.0 / −21.0	36.0 / −11.0
26.10.78	08.03.49	08.03.53	1.7×10^{-5}	253.340 / 60.210	59.0 ± 2.5						
2.11.78	12.33.30	12.33.32	2.6×10^{-6}	259.185 / 63.860	75.98 ± 0.09						
4.11.78a	4.30 ÷ 9.10	No data	5.5×10^{-6}								
4.11.78b	16.17.53	16.17.52	2.6×10^{-4}	81.091 / −64.820	86.86 ± 0.44						
7.11.78	22.10.27	No data	1.5×10^{-6}								
9.11.78	15.12.50	No data	2.0×10^{-6}								
11.11.78	17.22.02	Not detected	3.0×10^{-6}								
13.11.78	10.46.38	Not detected	4.5×10^{-6}								
15.11.78a	21.06.19	21.06.25	1.6×10^{-5}	273.400 / 69.330	33.1 ± 1.6						
15.11.78b	22.14.12	22.14.06	2.9×10^{-6}	93.440 / −69.360	26.7 ± 4.9						
17.11.78	13.46.42	No data	1.2×10^{-5}								
19.11.78	09.28.33	No data	3.2×10^{-4}								
21.11.78a	01.34.14	No data	9.0×10^{-5}								
21.11.78b	06.00.48	No data	7.0×10^{-6}								
23.11.78	16.06.48	16.06.45	3.0×10^{-6}	106.900 / −72.290	60.4 ± 2.5						
24.11.78	03.53.51	03.53.52	3.5×10^{-5}	288.000 / 72.480	75.3 ± 2.3	180.0 / 20.9	240.0 / 77.1	178.2 / 19.0	178.8 / 18.9	181.2 / 23.0	181.9 / 22.8
7.12.78	22.11.29	22.11.33	1.5×10^{-6}	339.190 / 75.069	47.8 ± 6.8	213.5 / 52.3	95.5 / 60.5	224.6 / 42.9	233.4 / 55.4	196.8 / 47.9	196.0 / 61.9
29.12.78	12.05.05	No data	2.0×10^{-6}			297.2 / −7.5	32.4 / −16.7	298.9 / −8.8	296.1 / −9.2	298.2 / −5.9	295.4 / −6.2
1.01.79	00.07.20	No data	5.2×10^{-5}			178.0 / 20.5	237.3 / 75.3	173.9 / 18.9	179.5 / 16.5	176.5 / 24.2	182.0 / 22.0
2.01.79a	17.55.31	No data	6.5×10^{-6}			101.9 / −76.0	286.6 / −26.6	104.5 / −69.8	76.1 / −75.9	124.0 / −73.6	95.6 / −82.2
2.01.79b	22.36.25	No data	1.7×10^{-6}								
3.01.79	5.10 ÷ 10.13	No data	2.3×10^{-7}								
7.01.79	05.31.42	No data	1.4×10^{-6}								
13.01.79	07.33.00	07.32.54	1.1×10^{-4}	217.502 / −21.492	57.5 ± 1.3	252.3 / −76.3	314.9 / −20.0	245.5 / −78.9	246.0 / −75.8	258.9 / −76.7	256.9 / −73.5

Table I (continued)

Date of event	T_0, h.m.s. UT		$S(E\gamma > 30\ \mathrm{keV})$, erg cm^{-2}	Triangulation ring		Mean source position		Box coordinates			
	V-11	V-12		Ring center α / δ	Ring radius	α / δ	l'' / b''	α / δ			
(1)	(2)	(3)	(4)	(5)	(6)	(7)	(8)	(9)			
16.01.79	08.59.18	08.59.15	2.3×10^{-5}	220.136 / −17.864	59.7 ± 1.6	169.0 / 14.3	238.7 / 64.7	166.0 / 13.3	170.2 / 11.6	167.7 / 17.1	172.0 / 15.4
19.01.79	14.22.35	14.22.31	2.0×10^{-5}	222.263 / −13.911	28.8 ± 2.8	246.3 / −32.6	346.4 / 11.0	251.9 / −29.8	245.2 / −28.9	247.2 / −36.3	239.2 / −35.0
31.01.79	10.26.14	No data	4.6×10^{-6}			190.4 / −66.7	301.6 / −4.1	183.9 / −70.8	182.4 / −64.6	199.9 / −68.4	195.2 / −62.3
8.02.79	10.05.29	10.05.28	2.3×10^{-6}	44.865 / −6.410	86.6 ± 4.3	327.4 / 48.7	94.5 / −4.0	323.6 / 58.3	320.6 / 48.0	334.4 / 49.0	330.1 / 38.9
11.02.79	11.43.25	11.43.24	4.0×10^{-7}	225.759 / 6.351	83.4 ± 1.0						
13.02.79	12.58.01	12.58.01	4.4×10^{-6}	226.094 / 6.248	83.0 ± 1.8	308.8 / −2.2	43.3 / −24.5	310.2 / −5.9	311.2 / 2.8	306.4 / −7.0	307.4 / 1.6
14.02.79	No data	12.00.46	4.8×10^{-6}								
15.02.79	17.14.56	17.14.48	3.3×10^{-5}	226.274 / 6.069	74.1 ± 3.5	299.7 / −4.4	36.5 / −17.6	302.8 / −7.8	303.9 / 0.8	295.4 / −9.5	296.6 / −1.0
24.02.79	21.50.59	No data	1.5×10^{-6}								
5.03.79	15.51.39	15.51.44	1.3×10^{-3}	44.428 / −2.747	68.522 ± 0.030	81.712 / −66.072	275.3 / −33.1	82.218 / −65.951	82.574 / −65.823	80.819 / −66.319	81.196 / −66.192
6.03.79	06.17.25	06.17.30	6.5×10^{-6}	44.341 / −2.591	68.679 ± 0.085	81.712 / −66.072	275.3 / −33.1	82.218 / −65.951	82.574 / −65.823	80.819 / −66.319	81.196 / −66.192
4.04.79	00.43.31	00.43.38	7.7×10^{-7}	43.708 / 6.015	77.142 ± 0.020	81.712 / −66.072	275.3 / −33.1	82.218 / −65.951	82.574 / −65.823	80.819 / −66.319	81.196 / −66.192
24.04.79b	18.27.18	18.27.25	4.0×10^{-7}	48.038 / 11.971	81.917 ± 0.012	81.712 / −66.072	275.3 / −33.1	82.218 / −65.951	82.574 / −65.823	80.819 / −66.319	81.196 / −66.192

Date	t_1	t_2	S	l / b	Δ	Coord 1	Coord 2	Coord 3	Coord 4	Coord 5	Coord 6
7.03.79	22.15.34	22.15.24	1.9×10^{-4}	224.108 / 2.148	50.58 ± 0.61	212.5 / −47.4	316.2 / 13.0	210.5 / −47.6	214.2 / −48.3	210.8 / −46.4	214.5 / −47.1
13.03.79	No data	17.26.16	6.3×10^{-5}								
14.03.79	No data	12.42.25	2.4×10^{-6}								
18.03.79	No data	23.28.54	1.2×10^{-6}								
23.03.79	23.40.07	23.39.11	3.4×10^{-5}			227.1 / −72.1	312.7 / −12.3	213.4 / −63.3	241.5 / −68.5	207.6 / −74.4	276.8 / −80.2
24.03.79	16.06.50	16.06.40	1.0×10^{-6}	222.997 / −2.825	64.016 ± 0.015	285.23 / 14.52	46.6 / 4.0	285.27 / 14.40	285.55 / 12.83	284.87 / 16.20	285.19 / 14.63
25.03.79a	01.58.19	01.58.09	1.5×10^{-6}	223.004 / −2.952	64.155 ± 0.025	285.23 / 14.52	46.6 / 4.0	285.27 / 14.40	285.55 / 12.83	284.87 / 16.20	285.19 / 14.63
27.03.79a	10.30.35	10.30.24	3.5×10^{-7}	223.080 / −3.682	64.200 ± 0.012	285.23 / 14.52	46.6 / 4.0	285.27 / 14.40	285.55 / 12.83	284.87 / 16.20	285.19 / 14.63
25.03.79b	13.41.12	13.40.59	4.8×10^{-5}	223.016 / −3.103	58.38 ± 0.28	273.34 / 30.12	56.5 / 20.4	274.46 / 28.67	272.86 / 31.61	273.77 / 28.64	272.12 / 31.59
27.03.79b	16.28.19	16.27.58	1.0×10^{-5}	223.091 / −3.759	34.7 ± 1.0						
29.03.79	22.28.42	22.28.35	6.5×10^{-5}			115.0 / 40.8	178.0 / 26.7	117.3 / 38.7	112.2 / 39.5	117.9 / 42.1	112.5 / 42.9
31.03.79	21.02.37	21.02.27	7.9×10^{-5}	223.377 / −5.052	68.54 ± 0.28	291.7 / 1.5	38.0 / −7.7	292.3 / −2.5	291.8 / −2.6	291.5 / 5.5	290.9 / 5.4
2.04.79a	01.27.07	01.27.06	2.3×10^{-6}	223.490 / −5.415	67.81 ± 0.27	291.75 / −7.33	30.1 / −11.8	292.2 / −18.4	291.0 / 3.7	291.6 / −18.5	290.5 / 3.6
2.04.79b	09.41.45	09.41.46	3.7×10^{-5}	43.525 / 5.520	85.7 ± 1.0	117.5 / −55.8	268.3 / −14.4	118.3 / −57.9	120.2 / −54.0	114.6 / −57.5	116.8 / −53.7
5.04.79	03.48.31	Not detected	3.0×10^{-7}								
6.04.79	11.41.08	11.41.16	1.0×10^{-6}	44.031 / 6.758	74.614 ± 0.020	345.29 / −47.01	342.2 / −61.2	351.95 / −53.85	352.00 / −53.82	340.31 / −39.41	340.34 / −39.37
12.04.79a	12.48.45	Not detected	6.0×10^{-6}	45.134 / 8.661	51.3 ± 1.0	93.3 / −9.3	216.7 / −12.1	95.0 / −7.5	93.7 / −11.2	92.9 / −7.4	91.5 / −11.3
12.04.79b	22.04.42	22.05.04	2.2×10^{-5}	45.685 / 9.419	76.819 ± 0.080						
15.04.79	13.07.55	13.08.04	1.6×10^{-6}	46.327 / 10.204	44.96 ± 0.10	326.76 / 16.43	71.7 / −27.9	326.04 / 22.68	327.44 / 10.15	326.21 / 22.67	327.60 / 10.13
18.04.79	07.44.41	07.45.10	6.5×10^{-5}			87.65 / −7.95	212.8 / −16.5	88.43 / −6.30	87.02 / −9.59	88.21 / −6.31	86.79 / −9.60

Table I (continued)

(1)	(2)	(3)	(4)	(5)	(6)	(7)	(8)	(9)	
Date of event	T_0, h.m.s. UT		$S(E\gamma >$ > 30 keV), erg cm^{-2}	Triangulation ring		Mean source position		Box coordinates	
	V-11	V-12		Ring center α δ	Ring radius	α δ	l'' b''	α δ	α δ
(1)	(2)	(3)	(4)	(5)	(6)	(7)	(8)	(9)	
19.04.79	16.05.02	16.06.23	4.1×10^{-4}			355.8 −38.8	348.3 −72.2	331.9 −37.7	4.6 −23.3	344.3 −55.3	22.6 −36.6
24.04.79a	10.39.42	Not detected	7.0×10^{-7}							
30.04.79	20.06.41	No data	3.8×10^{-6}			356.2 −17.3	63.5 −72.1	349.6 −15.0	357.9 −11.0	354.5 −23.4	3.1 −19.3
2.05.79	04.18.47	No data	2.3×10^{-5}			175.0 20.0	233.8 72.6	163.6 16.3	178.5 9.4	171.5 30.6	187.4 23.0
4.05.79	08.38.52	No data	5.7×10^{-6}			345.0 32.0	96.9 −25.2	339.8 20.2	355.6 28.1	333.7 35.6	352.4 45.6
14.05.79	17.51.02	No data	6.5×10^{-6}			37.6 53.5	137.2 −6.2	44.9 52.4	36.2 49.8	39.1 57.2	29.6 54.2
15.05.79	14.30.16	No data	1.7×10^{-6}			63.0 −13.8	207.0 −40.9	69.5 −4.4	54.9 −8.6	71.4 −19.2	55.5 −24.2
19.05.79	11.40.41	No data	3.6×10^{-6}			358.0 23.0	105.7 37.8	352.5 9.2	11.1 18.0	344.2 27.6	5.3 38.0
24.05.79	03.56.26	No data	1.5×10^{-5}							

A. THE DISTRIBUTION OF BURSTS IN DURATION

The essential differences in the gamma-burst time structure are reflected in the distribution of the observed events in duration T_b. Figure 164 shows an experimental distribution drawn for 143 events. It displays the number of bursts per equal logarithmic interval of T_b. Since some of the bursts may have long tails, the duration of the event in this case is taken to be the interval of time within which fall 80–90% of the measured burst intensity S. The distribution differs substantially from the uniform one. The main peak in the distribution is connected primarily with single and multipulse bursts. The right-hand wing is composed of double and long structureless bursts. Narrow peak in the beginning of the graph indicates the existence of a separate class of short bursts.

B. DISTRIBUTION OF GAMMA-BURST SOURCES OVER THE CELESTIAL SPHERE

Figure 165 shows the gamma-burst source distribution over the celestial sphere in galactic coordinates. There are presented locations of 58 sources determined in the KONUS experiment and of 11 sources from the review of Hurley (1979), determined in earlier experiments. We see that these locations are distributed in a random way over the sky and do not reveal any sign of anisotropy. Our earlier data (Mazets and Golenetskii, 1981), based on a smaller number of sources, seemed to indicate a weak concentration of the sources toward the galactic disc plane and to direction to the galactic center. However, as the amount of experimental data has increased and the statistical accuracy improved, this trend has disappeared. Figure 166 shows the source distribution in galactic latitude. The histogram gives the number of sources with absolute galactic latitude above a given value. The smooth curve displays the relation calculated for an isotropic distribution. The experimental and the theoretical curves are seen to fit closely. The next graph (Figure 167) gives the number of sources with an angular distance from the direction to the galactic center as more than φ. We again see that the experimental curve fits excellently to the relation for an isotropic distribution. This graph also stresses the equivalence between the celestial hemisphere facing the galactic center and the opposite hemisphere.

C. LOG N – LOG S AND LOG N – LOG P DISTRIBUTIONS

The average frequency of gamma-burst detection in the KONUS experiment at the sensitivity threshold $S_{min} \approx 5 \times 10^{-7}$ erg cm^{-2} (for $E\gamma > 30$ keV) is ≈ 0.4 events day^{-1}. The dependence of the gamma-burst occurence frequency $N(>S)$ on their intensity S, is shown in Figure 168 on a log-log scale. This distribution deviates drastically from the $S^{-3/2}$ law; its slope is less than unity. Figure 169 displays the $\log N(>P) - \log P$ distribution for the burst occurrence frequency vs. measured radiation power at the highest burst peak (for bursts longer than 1 s). The initial part of the curve obeys the $P^{-3/2}$ law transferring after that to the P^{-1} distribution.

D. DISCUSSION

Isotropic distribution of the gamma-burst sources over the sky and the flat $\log N - \log S$ plot seem to be contradictory. This contradiction, however, is in no way real. In contrast to stationary sources, for instance, radio sources, the shape of the $\log N - \log S$ observational curve for pulsed gamma bursts depends, in a very complex manner, on the spatial source distribution, the spread in energy release, distribution of gamma bursts in duration, and the associated instrumental effects. An analysis of the data (Mazets *et al.*, 1980), in the first place, shows that the assumption of the average energy released in the sources being constant is not supported. The average burst radiation power appears to be more constant. The total energy of a burst turns out to be proportional to its duration. Radiation power for the peak of burst exhibits a larger spread. Nevertheless, even in the distribution of the burst occurrence frequency vs. peak radiation power (Figure 169) the part of the curve obeying the $P^{-3/2}$ law manifests itself clearly. Estimates show the data on the isotropic source distribution over the celestial sphere and the shape of the $\log N - \log P_{av}$ plot to agree well if one assumes the average burst energy to be $Q \sim 10^{38} T_b$ (erg), and the average distance to the observed sources to lie in the range 300 pc–2 kpc.

Acknowledgments

The KONUS experiment has become possible due to continuous assistance and support on the part of many organizations and specialists. The authors are deeply grateful to R. Z. Sagdeev, I. S. Shklovskii, A. S. Melioranskii, R. A. Sunyaev, N. F. Borodin, E. M. Vasilyev, V. I. Dvoretskii, F. I. Dol-gopolitcheskii, G. I. Zabiyaking, V. G. Zolotukhin, V. G. Kurt, E. V. Larionov, O. A. Magarshak, F. F. Mikhailus, G. S. Narimanov, D. P. Nikitinskaya, E. G. Pankov, V. M. Pokras, I. D. Skobkin, I. V. Sobatchkin, V. I. Subbotin, and P. E. Elyasberg for their helpful assistance in the conduction of the experiment and information processing, as well as for valuable discussions in the course of analysis of the data obtained.

References

Hurley, K.: 1979, *Adv. Space Exploration* **7**, 123.
Mazets, E. P. and Golenetskii, S. V.: 1981, *Astrophys. Space Sci.* **75**, 47.
Mazets, E. P., Golenetskii, S. V., Il'inskii, V. N., Panov, V. N., Aptekar, R. L., Gur'yan, Yu. A., Sokolov, I. A., Sokolova, Z. Ya., and Kharitonova, T. V.: 1979a, 'Catalog of Cosmic Gamma-ray Bursts from the KONUS Experiment Data', Part 1, A. F. Ioffe Institute preprint No. 618, Leningrad.
Mazets, E. P., Golenetskii, S. V., Il'inskii, V. N., Panov, V. N., Aptekar, R. L., Gur'yan, Yu. A., Sokolov, I. A., Sokolova, Z. Ya., and Kharitonova, T. V.: 1979b, 'Catalog of Cosmic Gamma-ray Bursts from the KONUS Experiment Data', Part 2, A. F. Ioffe Institute preprint No. 637, Leningrad.

Mazets, E. P., Golenetskii, S. V., Il'inskii, V. N., Panov, V. N., Aptekar, R. L., Gur'yan, Yu. A., Sokolov, I. A., Sokolova, Z. Ya. and Kharitonova, T. V.: 1979c, *Kosmitcheskiye Issledovaniya* **17**, 812.

Mazets, E. P., Golenetskii, S. V., Il'inskii, V. N., Panov, V. N., Aptekar, R. L., Gur'yan, Yu. A., Sokolov, I. A., Sokolova, Z. Ya., and Kharitonova, T. V.: 1979d, *Pis'ma v Astronomitcheskii Zhurnal* **5**, 313.

Mazets, E. P., Golenetskii, S. V., Aptekar, R. L., Gur'yan, Yu. A., and Il'inskii, V. N.: 1980, *Pis'ma v Astronomitcheskii Zhurnal* **6**, 609.

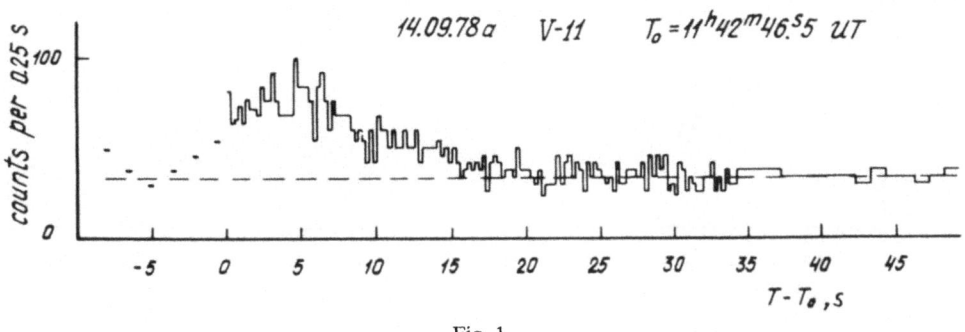

Fig. 1

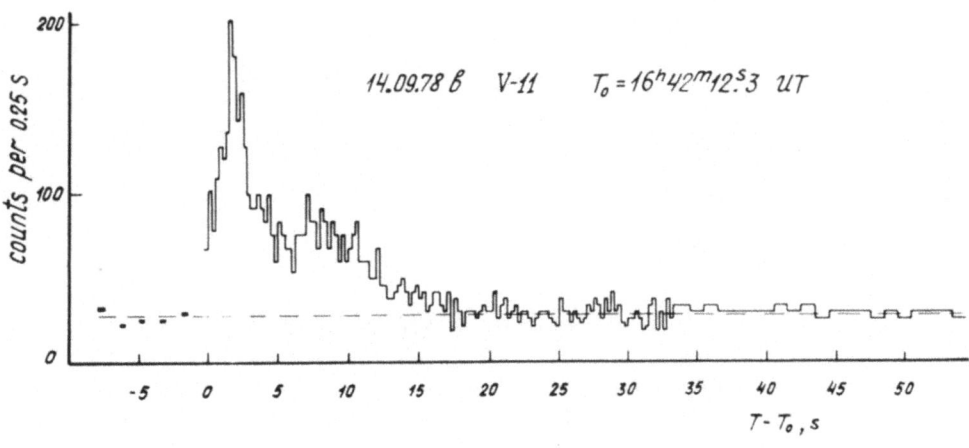

Fig. 2

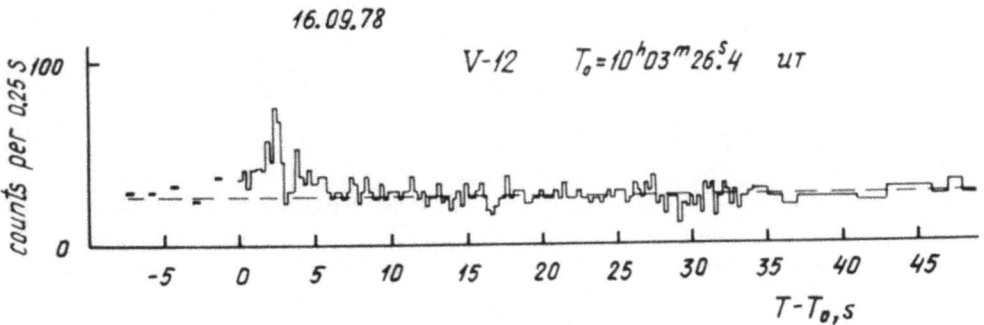

Fig. 3

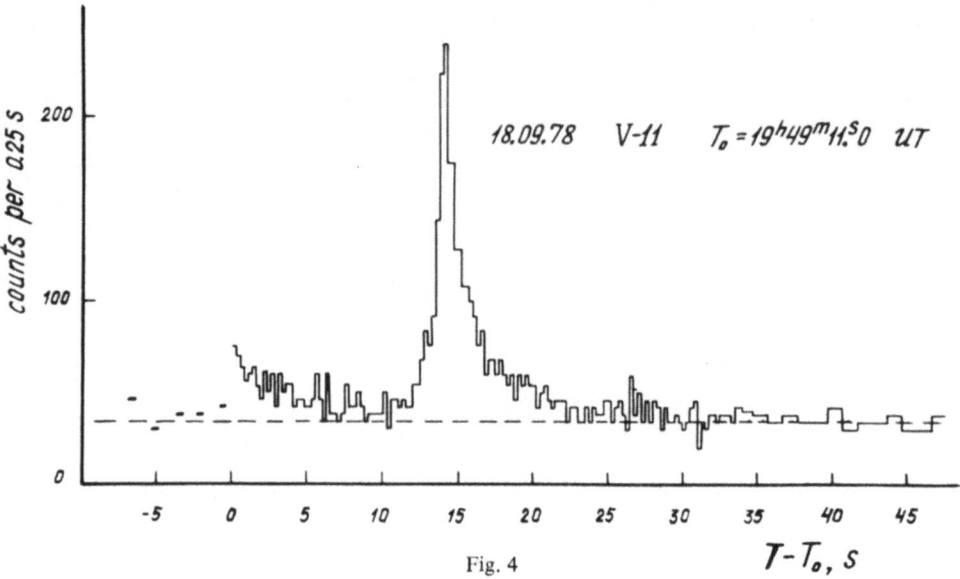

Fig. 4

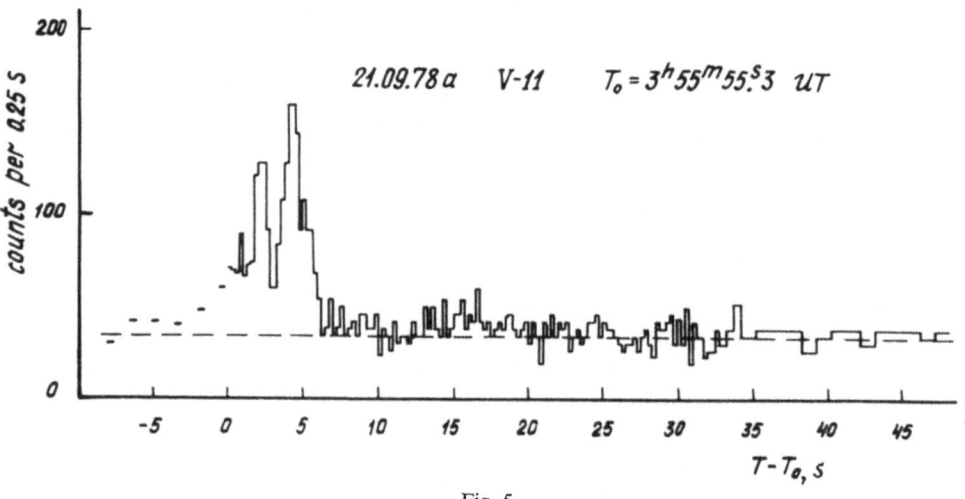

Fig. 5

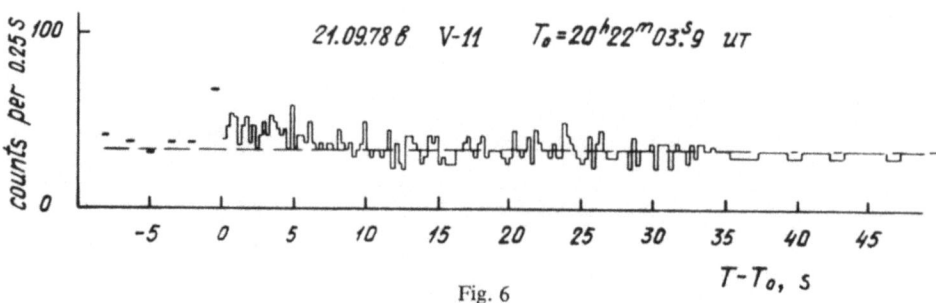

Fig. 6

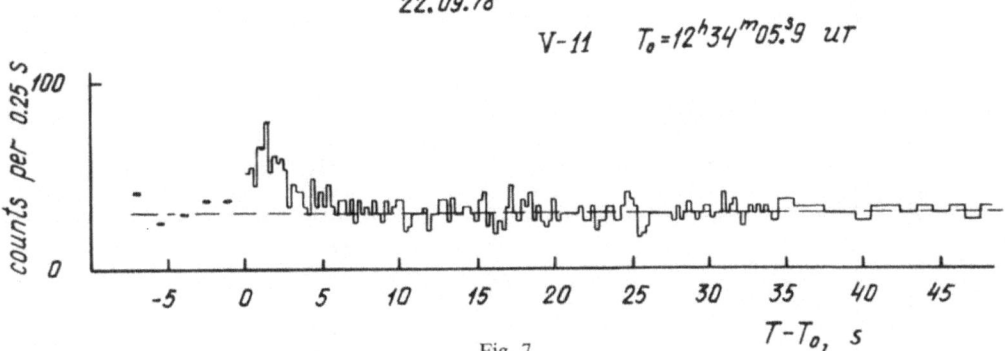

Fig. 7

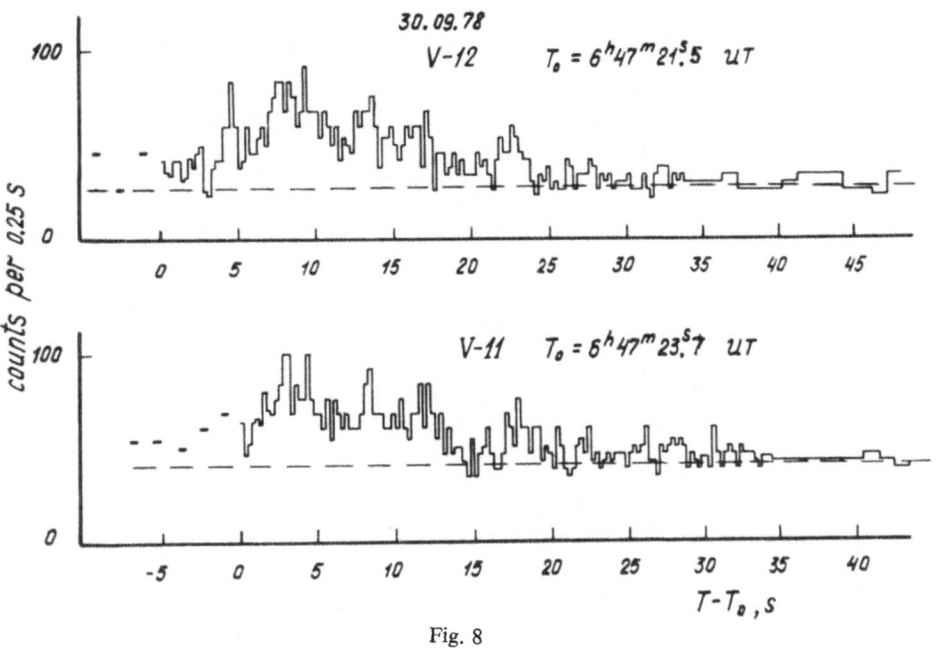

Fig. 8

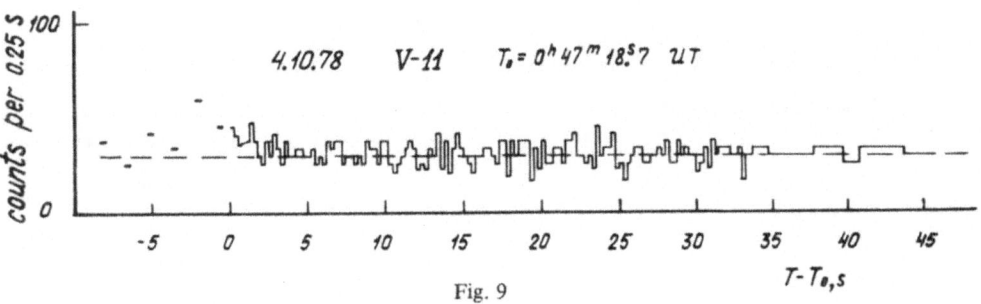

Fig. 9

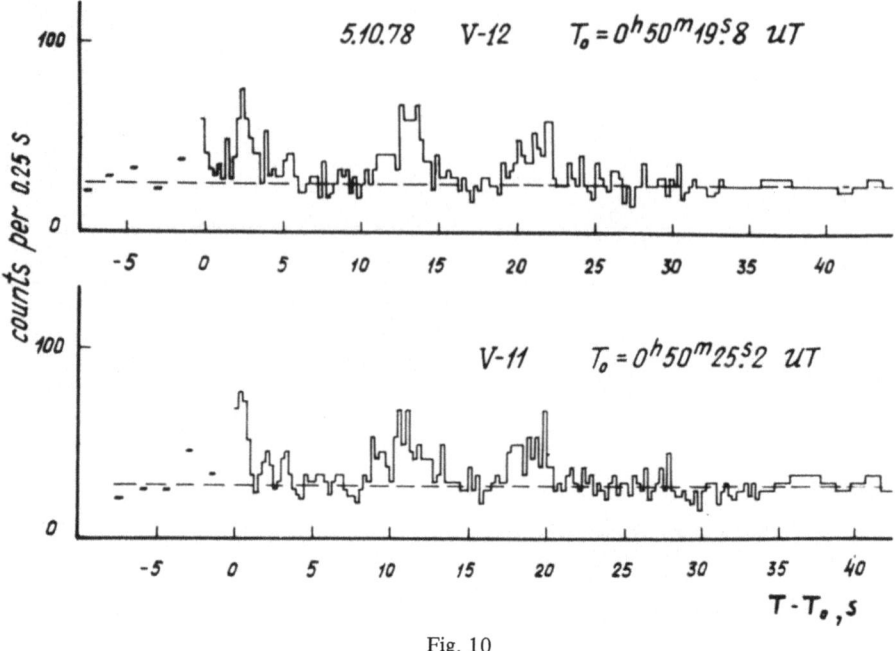

Fig. 10

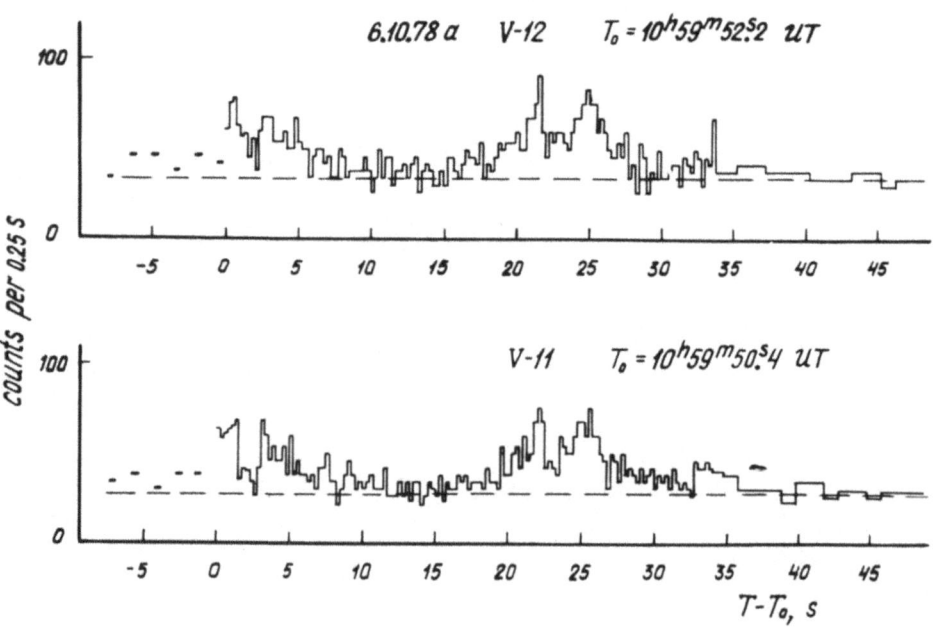

Fig. 11

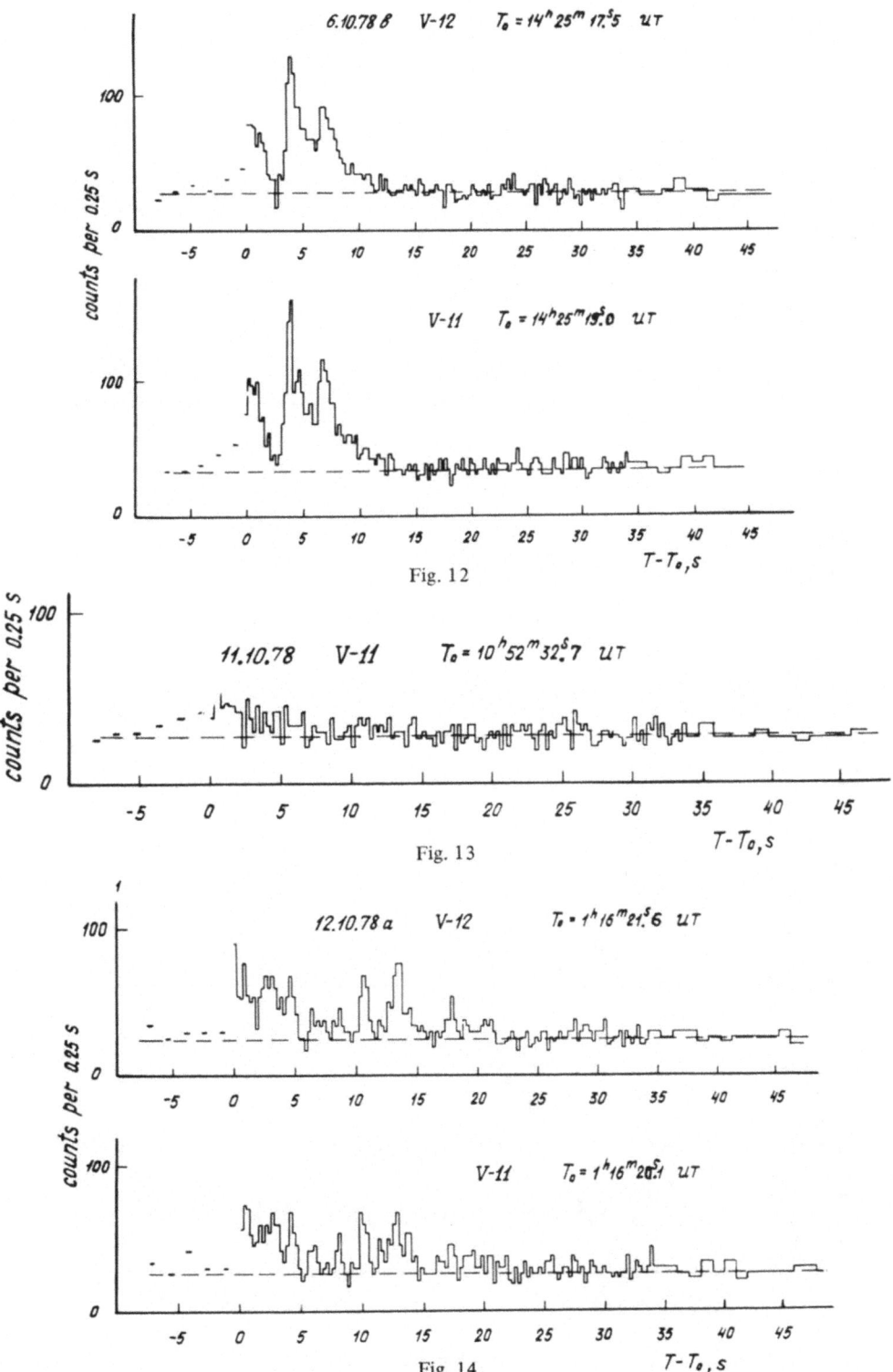

Fig. 12

Fig. 13

Fig. 14

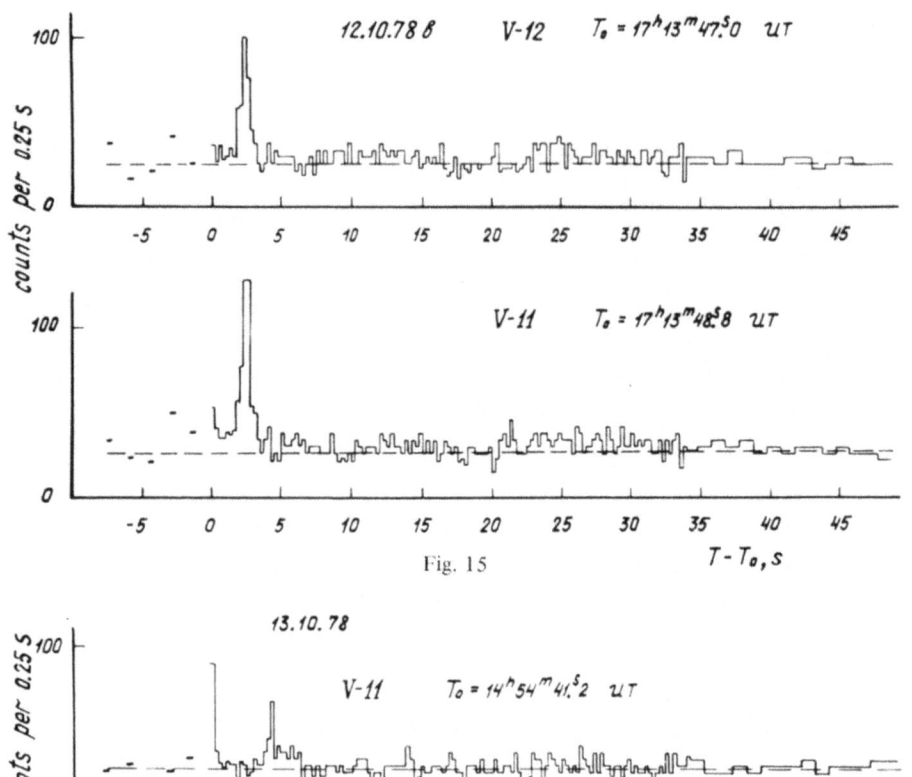

Fig. 15

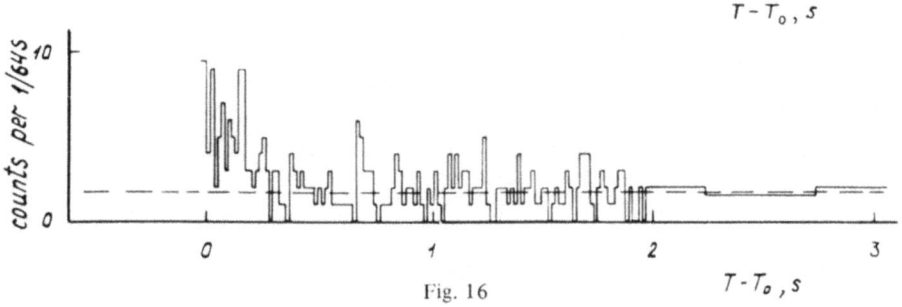

Fig. 16

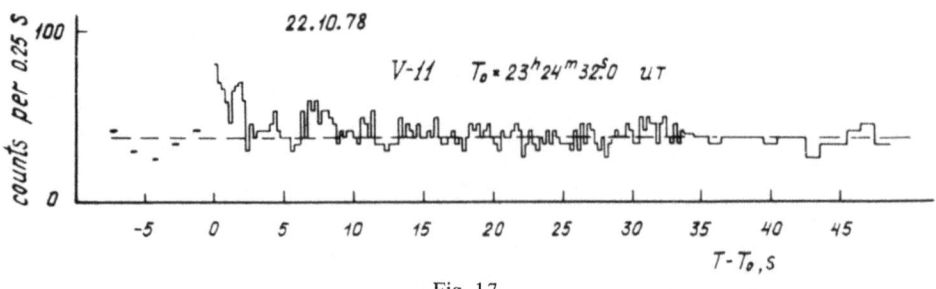

Fig. 17

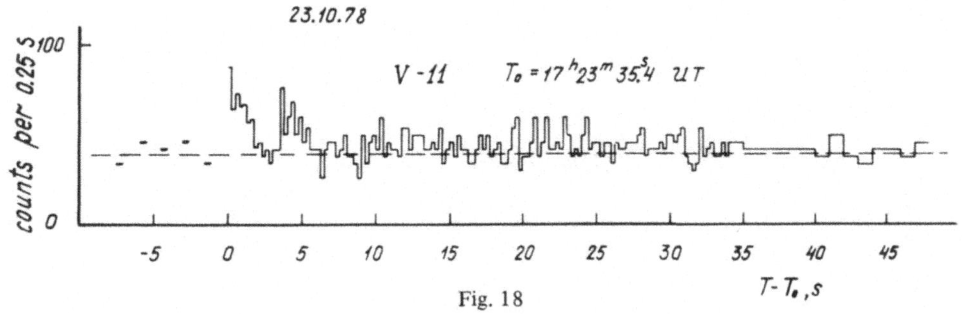

23.10.78

$V-11$ $T_0 = 17^h 23^m 35.4^s$ UT

counts per 0.25 s

100

0

-5   0   5   10   15   20   25   30   35   40   45

$T - T_0, s$

Fig. 18

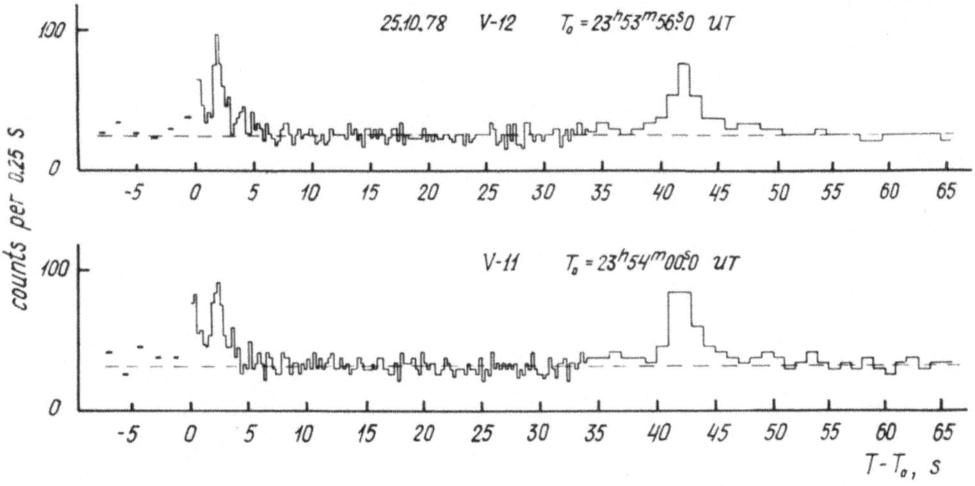

25.10.78 $V-12$ $T_0 = 23^h 53^m 56.0^s$ UT

counts per 0.25 s

100

0

-5   0   5   10   15   20   25   30   35   40   45   50   55   60   65

$V-11$ $T_0 = 23^h 54^m 00.0^s$ UT

100

0

-5   0   5   10   15   20   25   30   35   40   45   50   55   60   65

$T - T_0, s$

Fig. 19

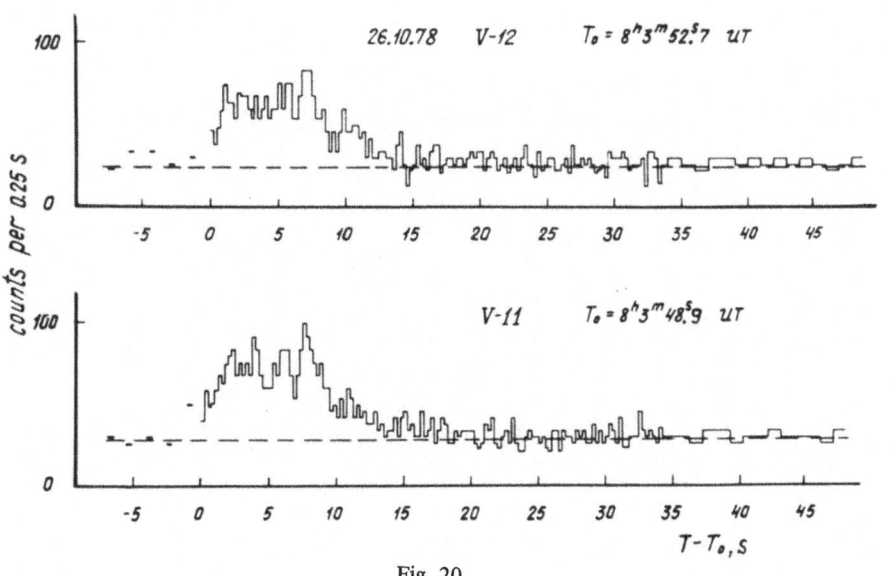

26.10.78 $V-12$ $T_0 = 8^h 3^m 52.7^s$ UT

counts per 0.25 s

100

-5   0   5   10   15   20   25   30   35   40   45

$V-11$ $T_0 = 8^h 3^m 48.9^s$ UT

100

0

-5   0   5   10   15   20   25   30   35   40   45

$T - T_0, s$

Fig. 20

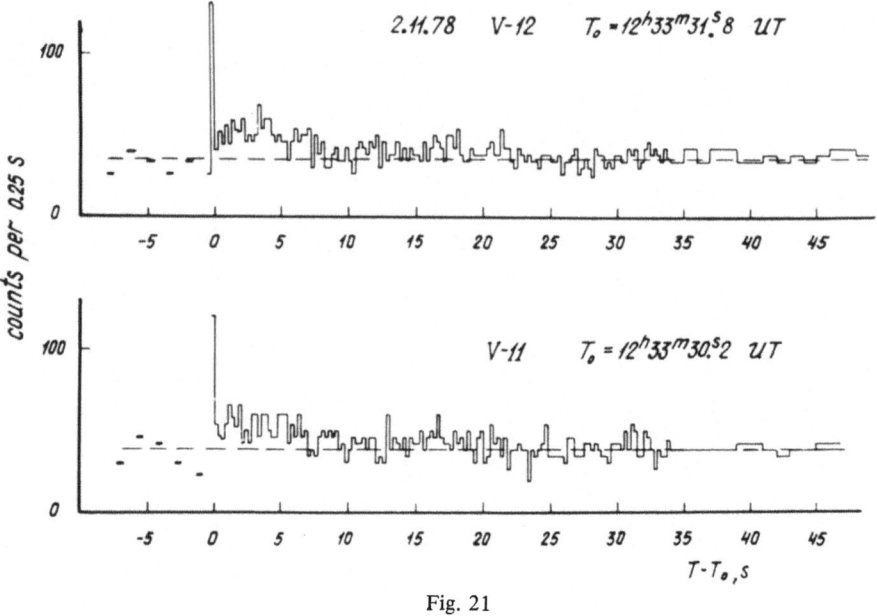

Fig. 21

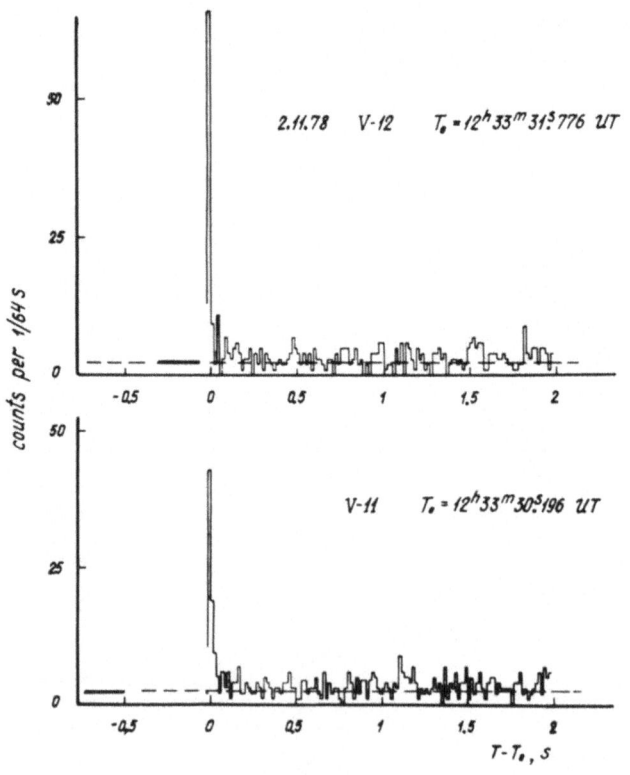

Fig. 22

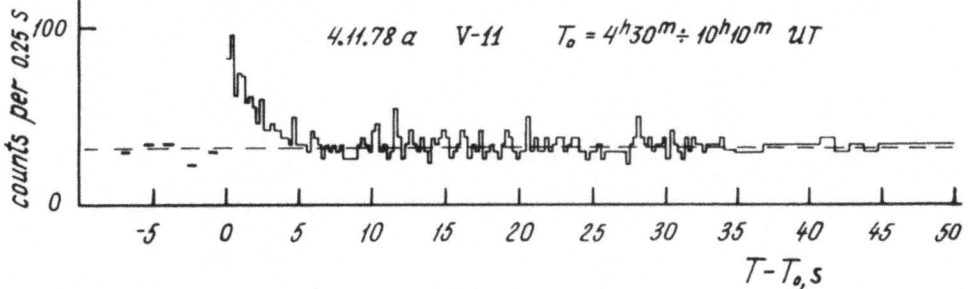

Fig. 23

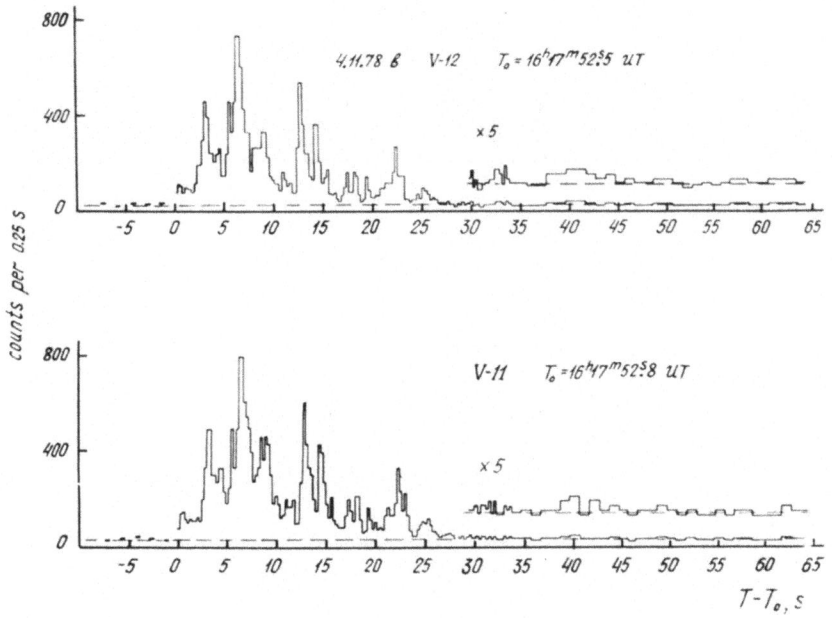

Fig. 24

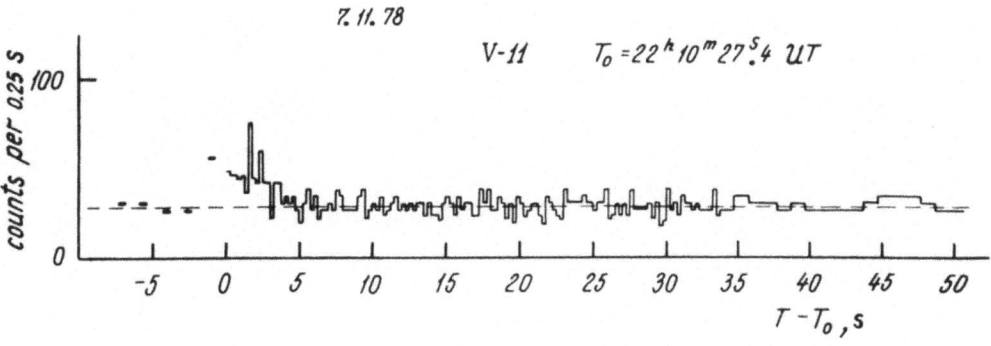

Fig. 25

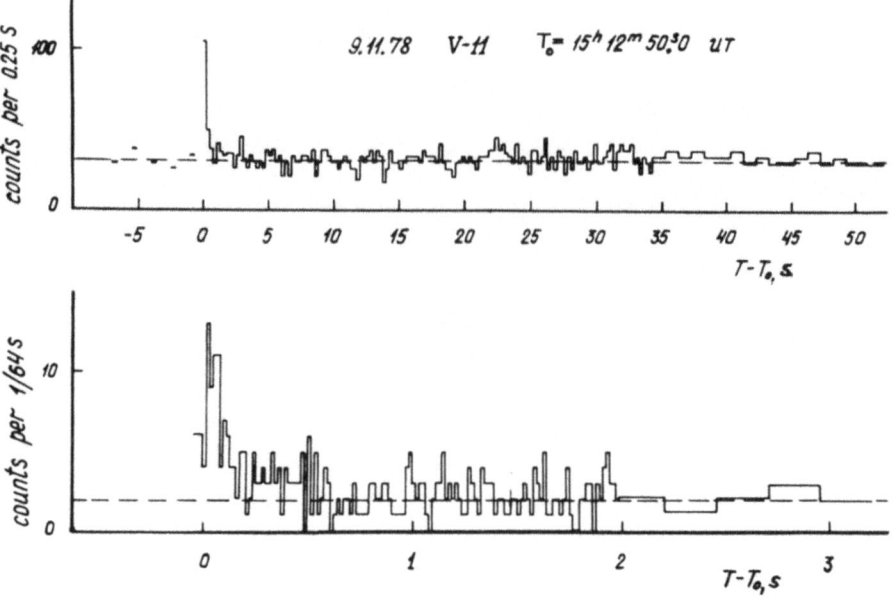

Fig. 26

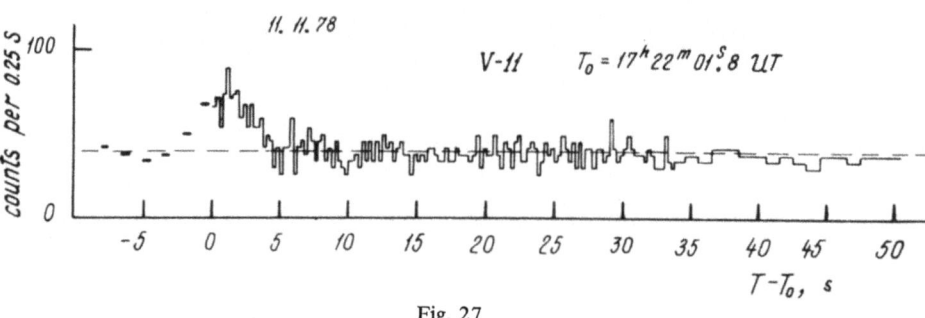

Fig. 27

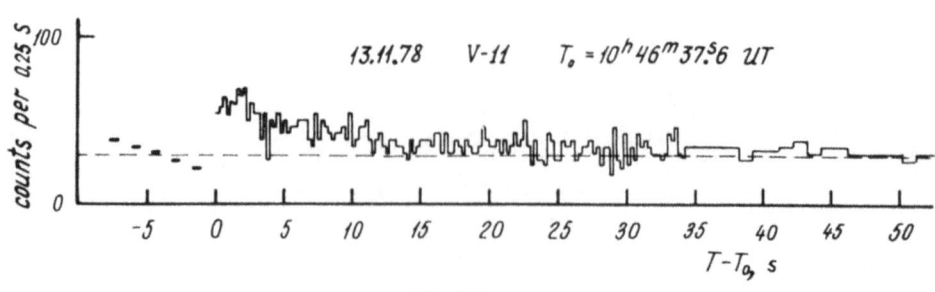

Fig. 28

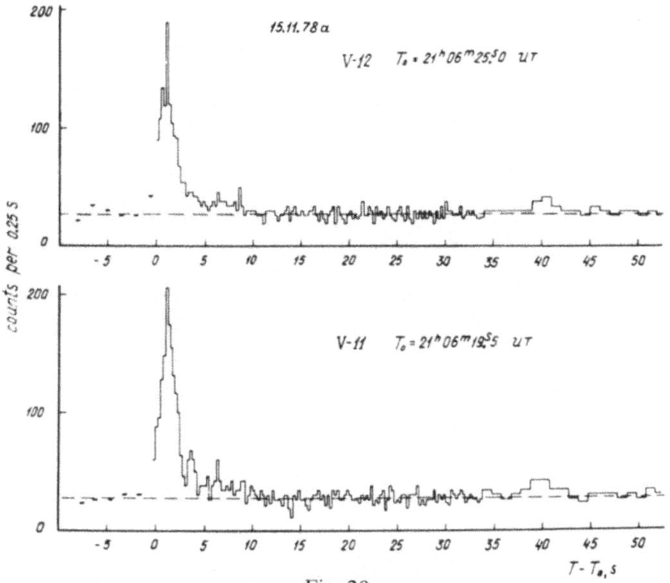

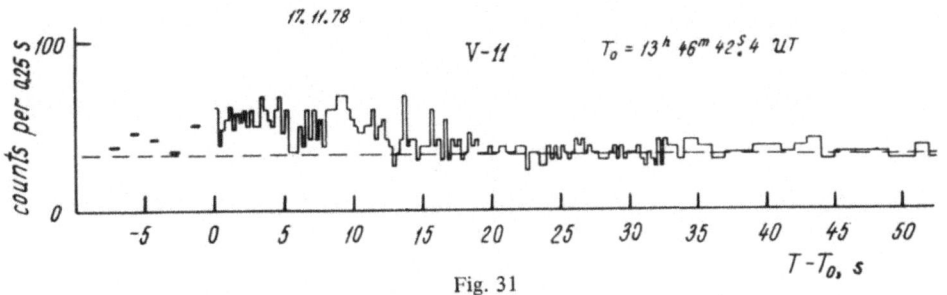

15.11.78 a

V-12 $T_0 = 21^h 06^m 25.^s0$ UT

V-11 $T_0 = 21^h 06^m 12.^s5$ UT

counts per 0.25 s

$T - T_0$, s

Fig. 29

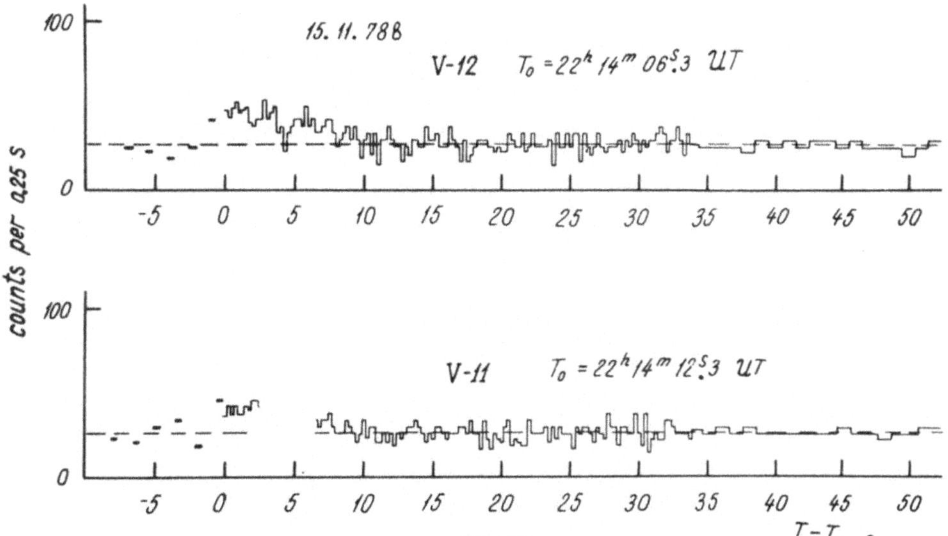

15. 11. 78 B

V-12 $T_0 = 22^h 14^m 06.^s3$ UT

V-11 $T_0 = 22^h 14^m 12.^s3$ UT

counts per 0.25 s

$T - T_0$, s

Fig. 30

17. 11. 78

V-11 $T_0 = 13^h 46^m 42.^s4$ UT

counts per 0.25 s

$T - T_0$, s

Fig. 31

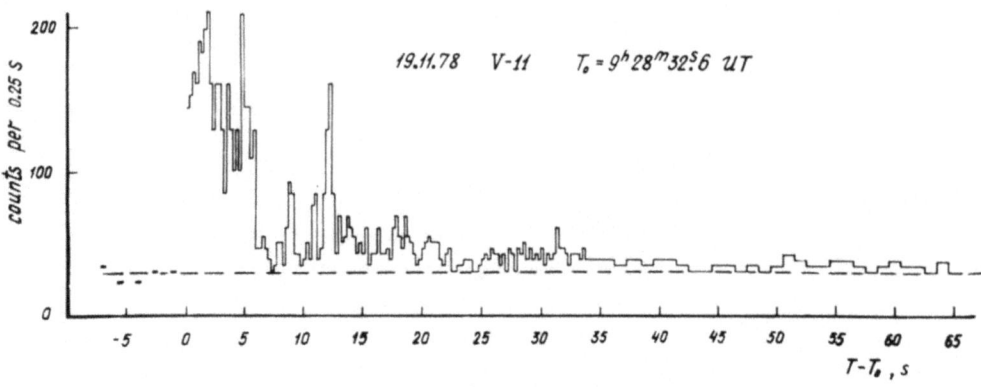

Fig. 32

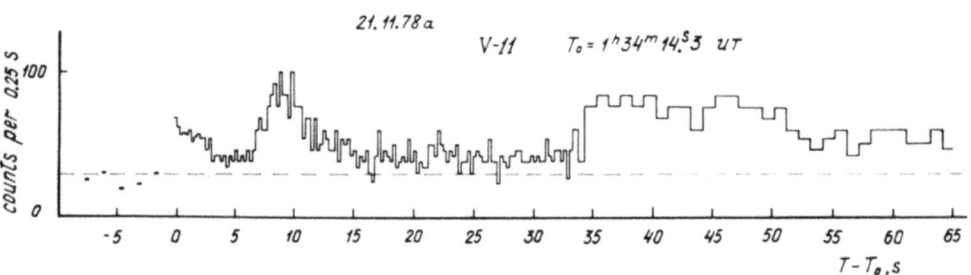

Fig. 33

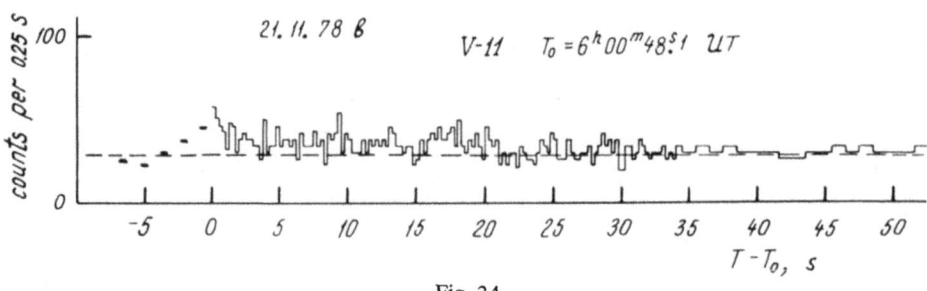

Fig. 34

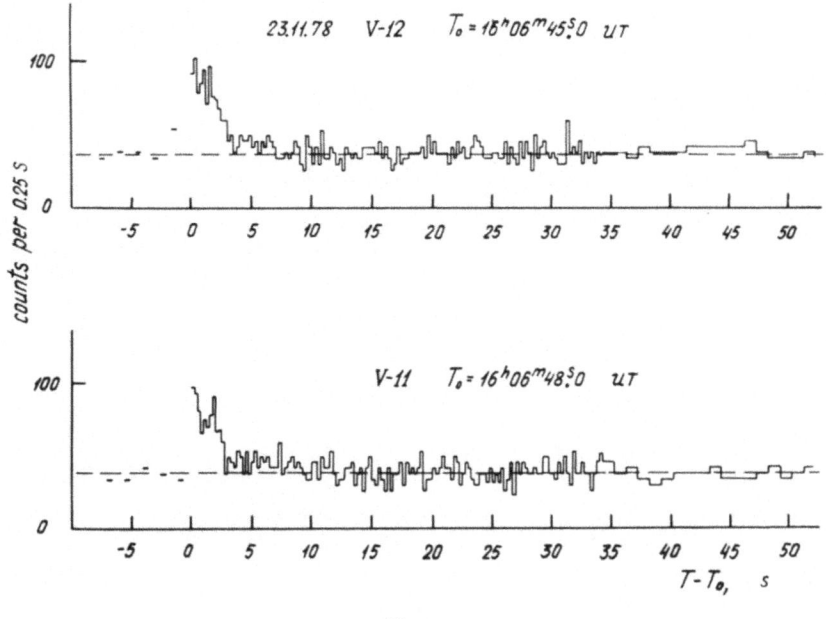

Fig. 35

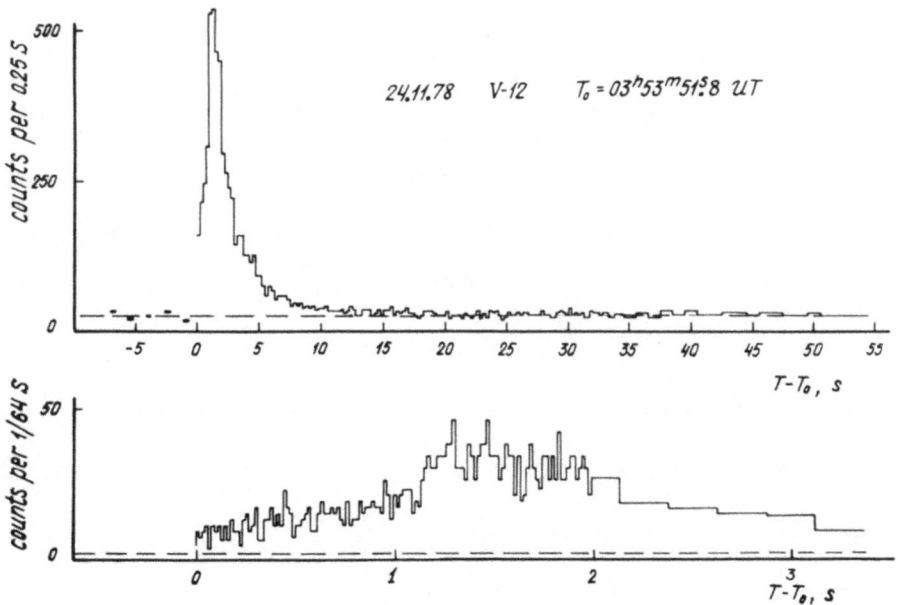

Fig. 36

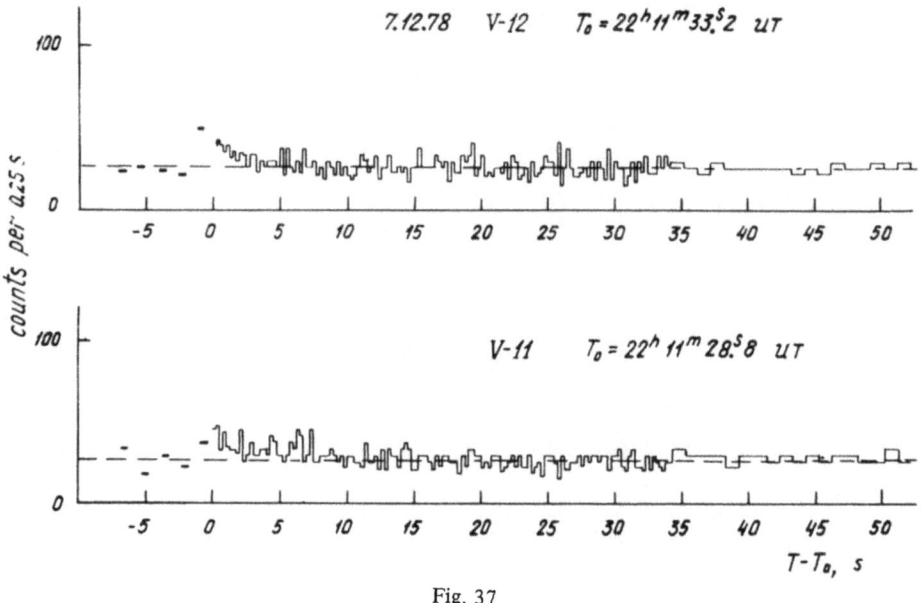

Fig. 37

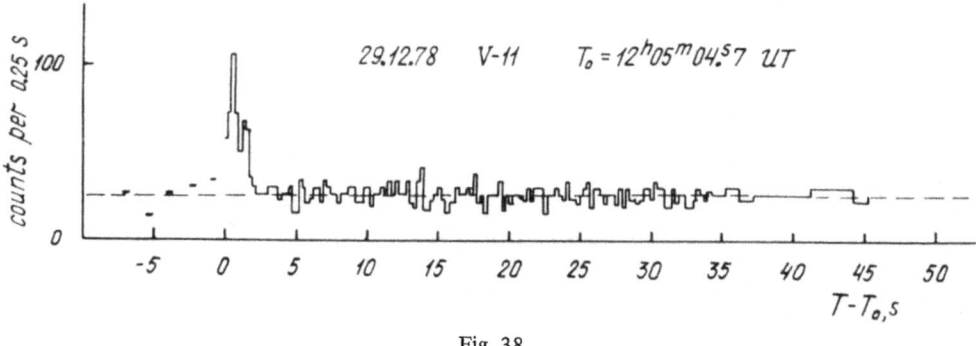

Fig. 38

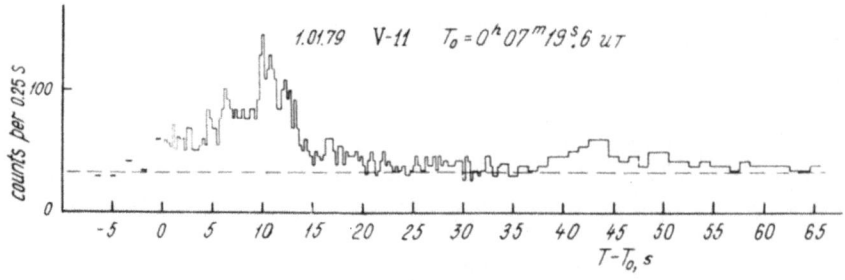

Fig. 39

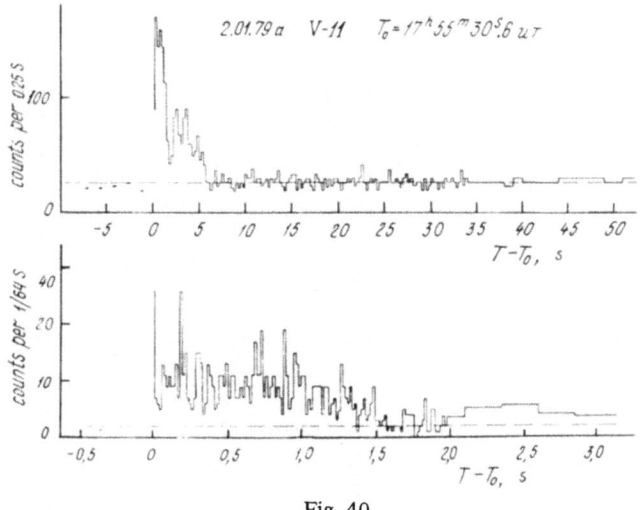

Fig. 40

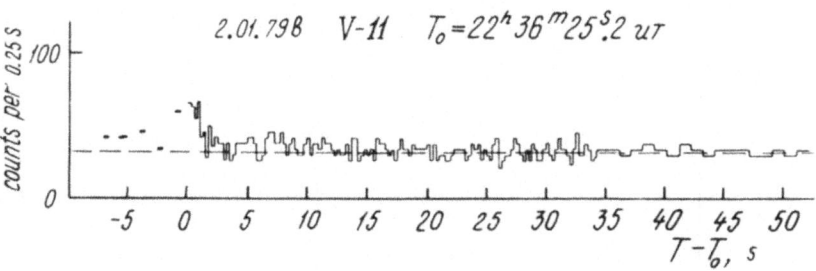

Fig. 41

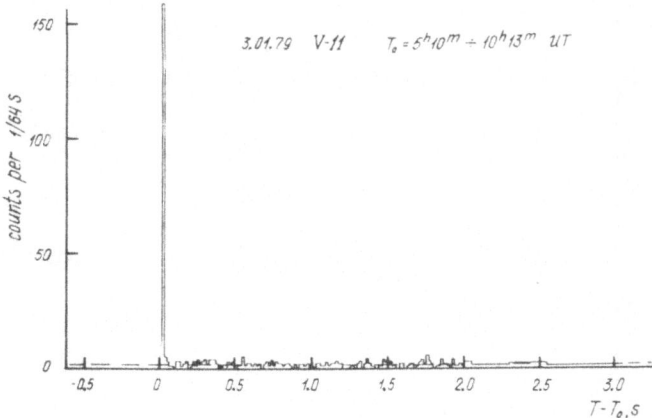

Fig. 42

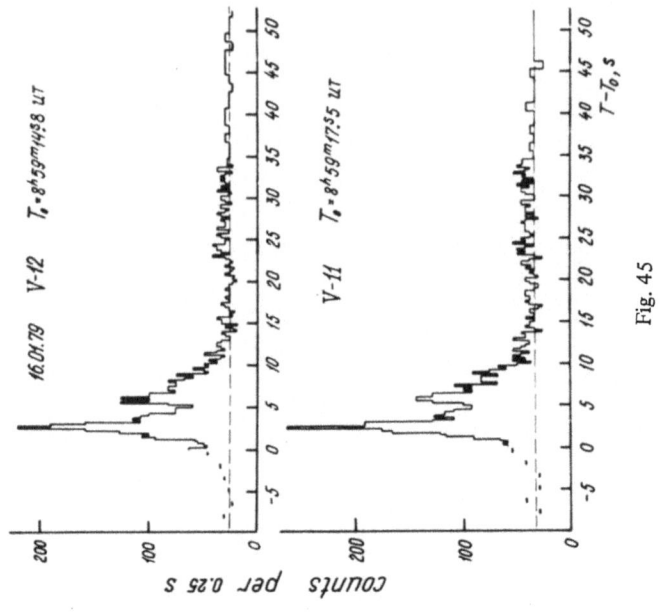

Fig. 45

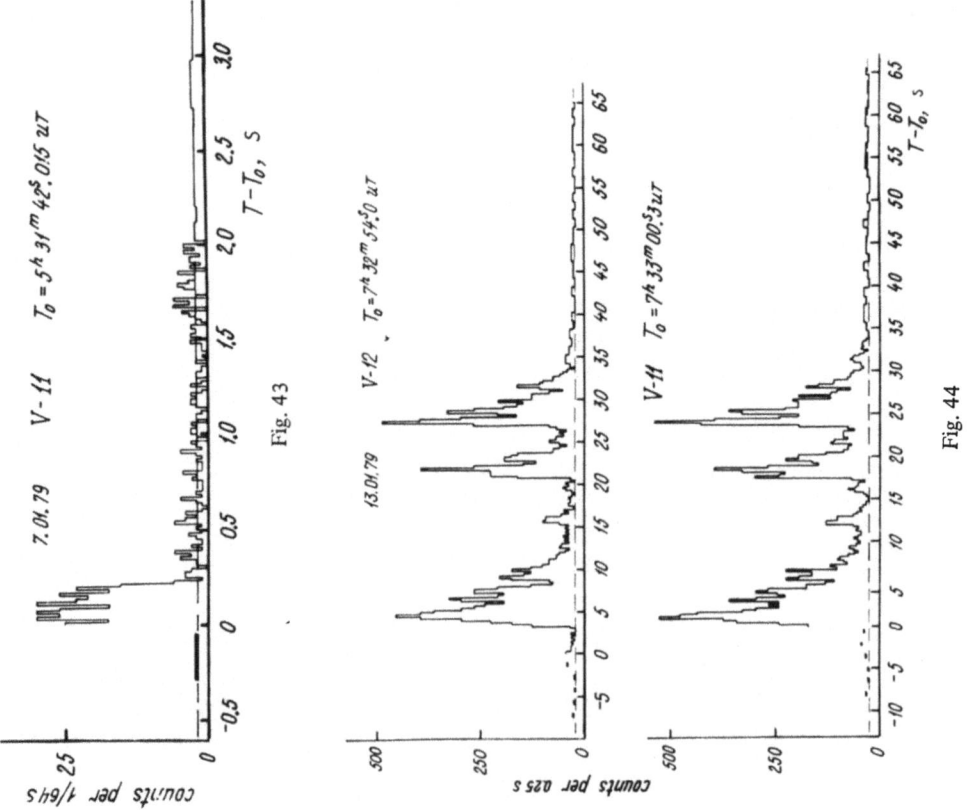

Fig. 43

Fig. 44

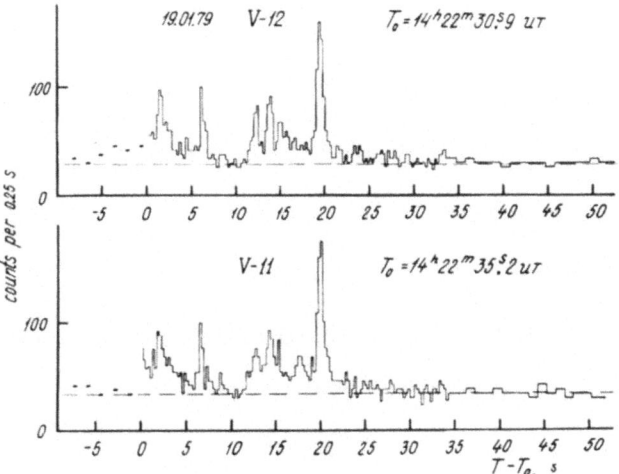

Fig. 46

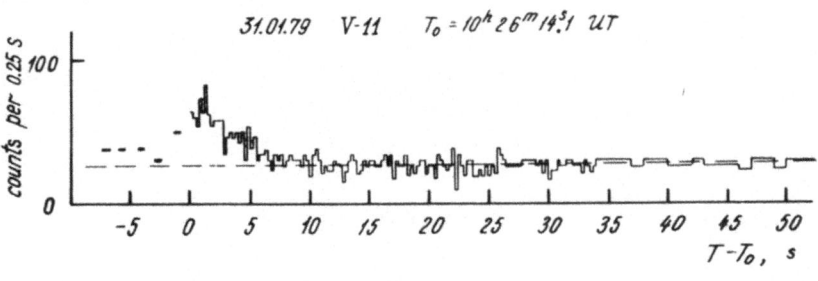

Fig. 47

Fig. 48

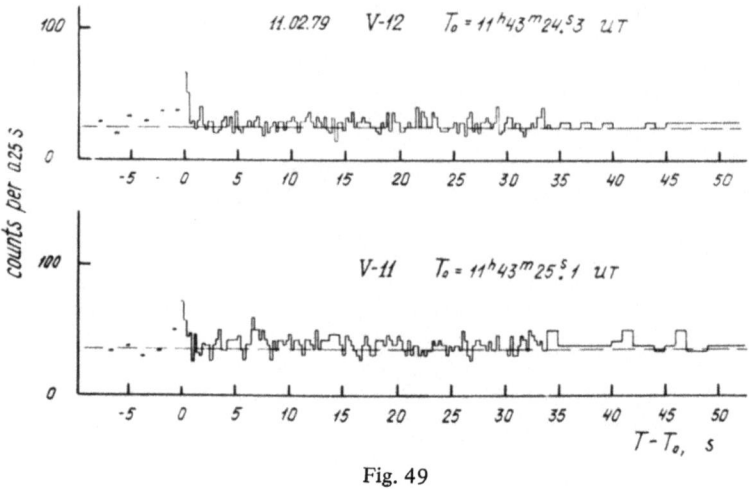

Fig. 49

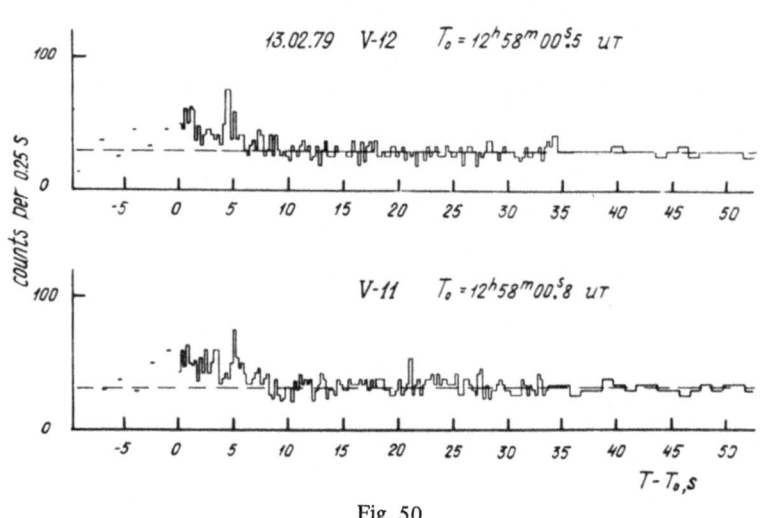

Fig. 50

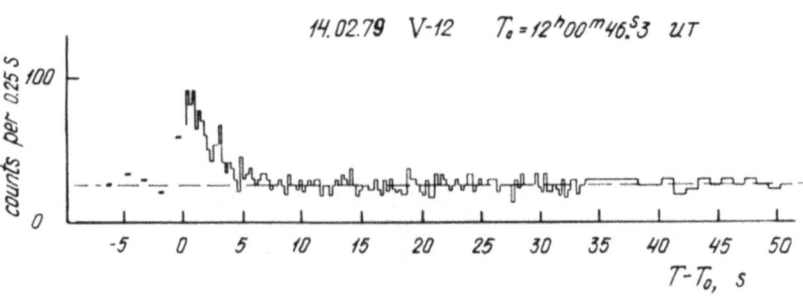

Fig. 51

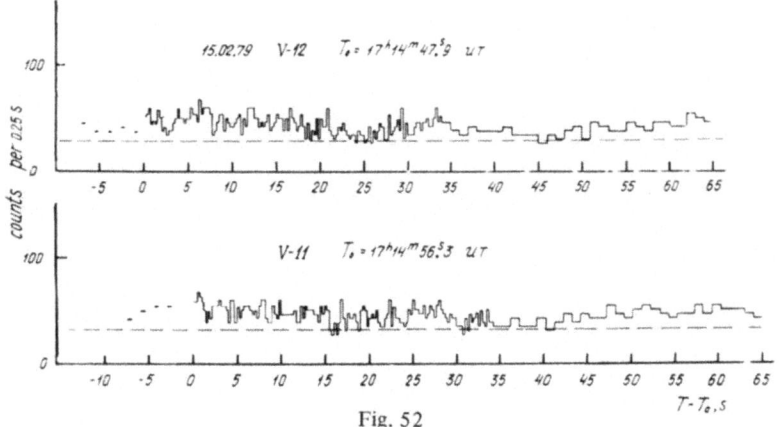

Fig. 52

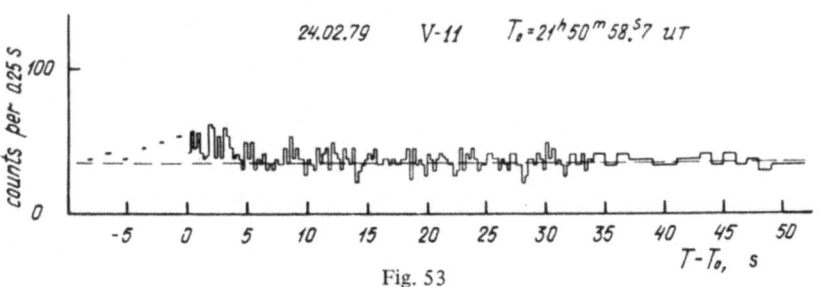

Fig. 53

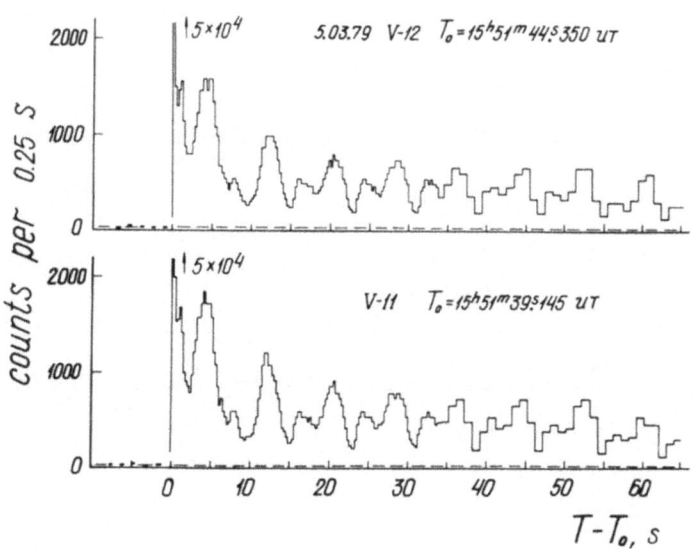

Fig. 54

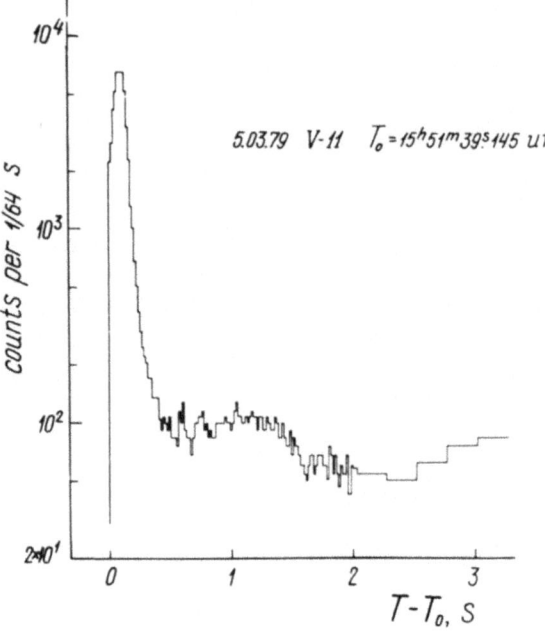

Fig. 55

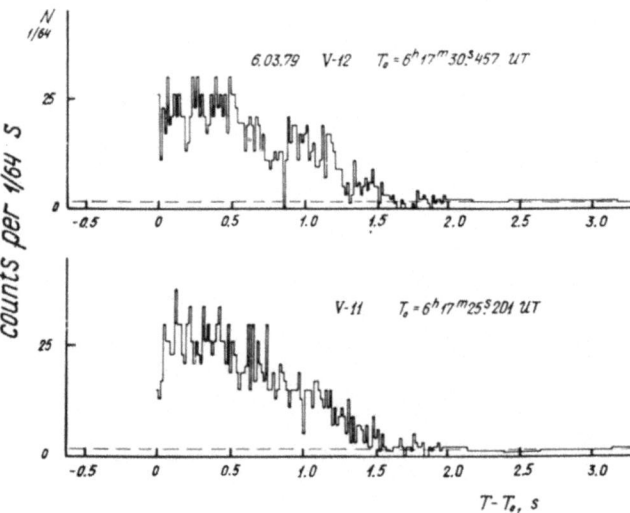

Fig. 56

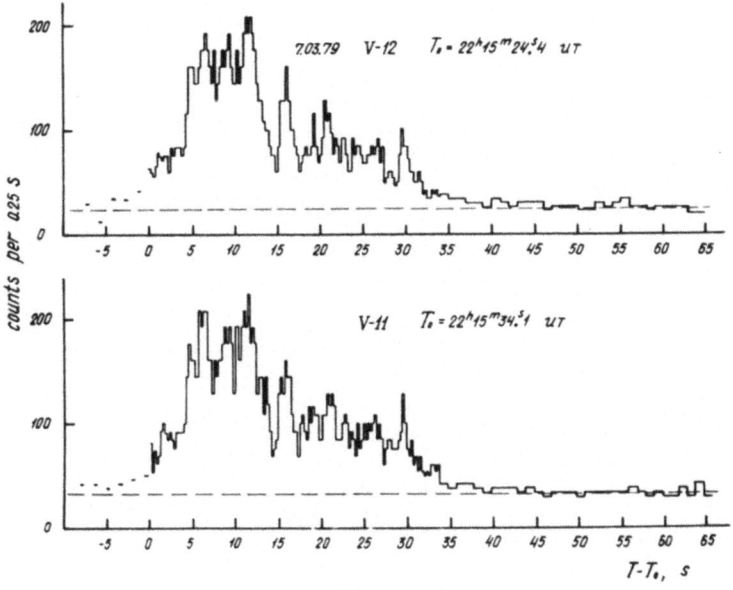

Fig. 57

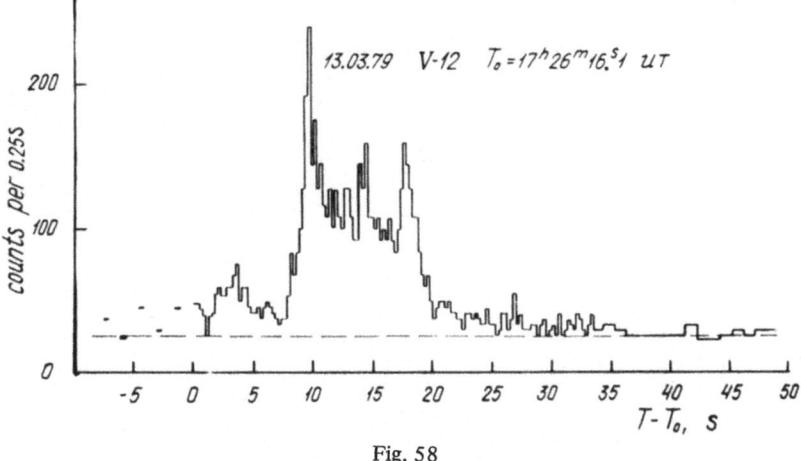

Fig. 58

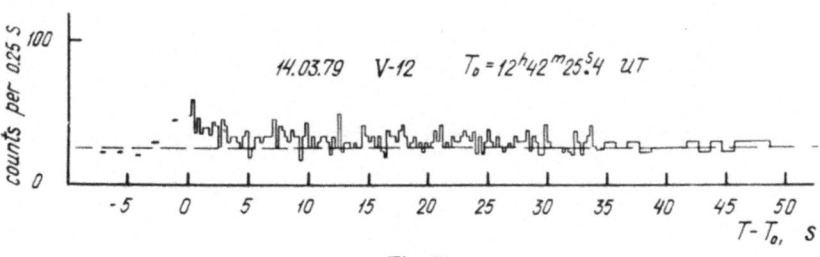

Fig. 59

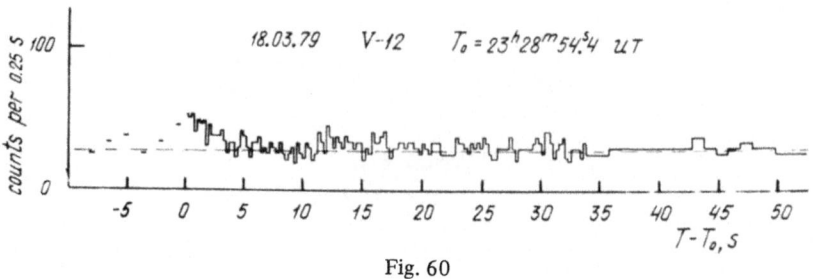

Fig. 60

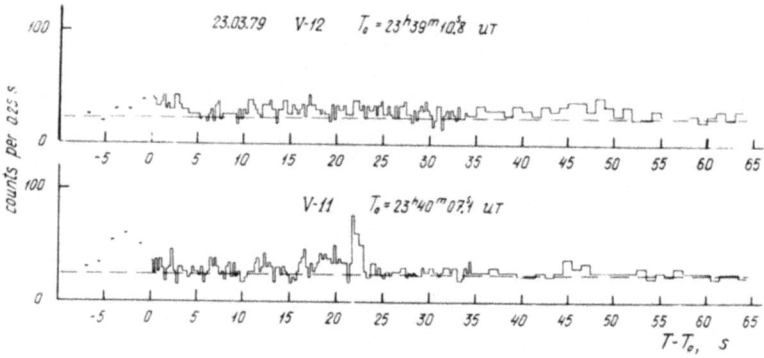

Fig. 61

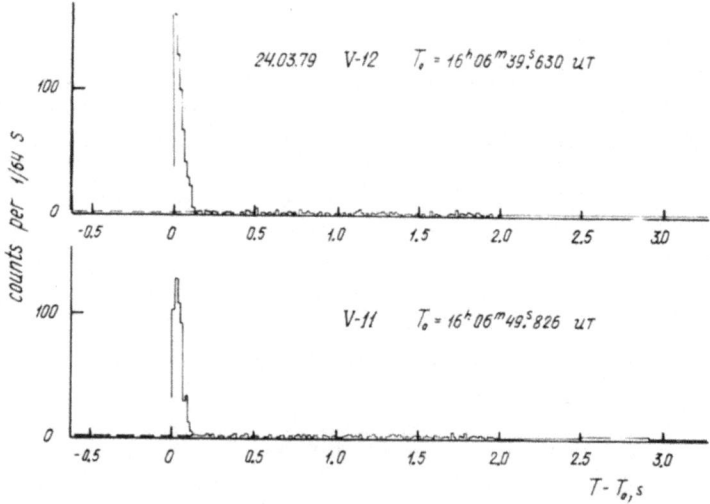

Fig. 62

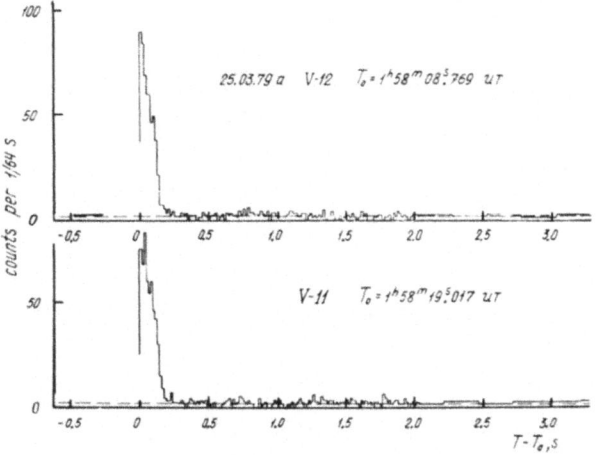

Fig. 63

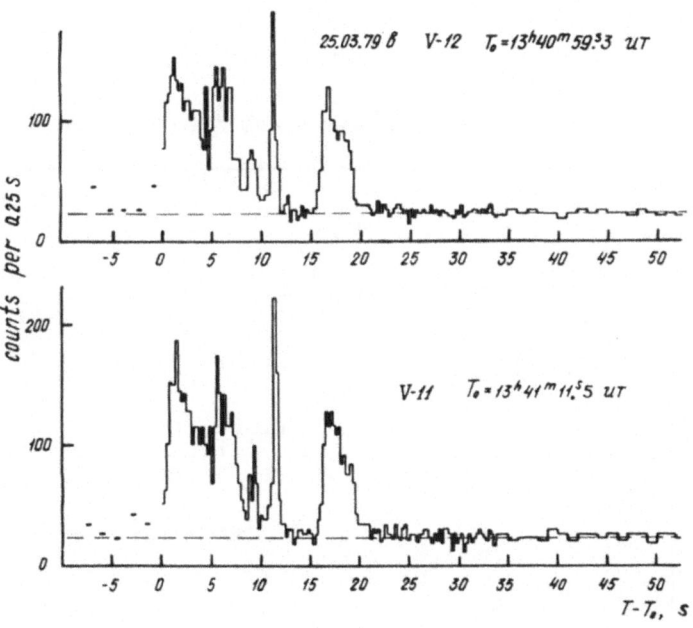

Fig. 64

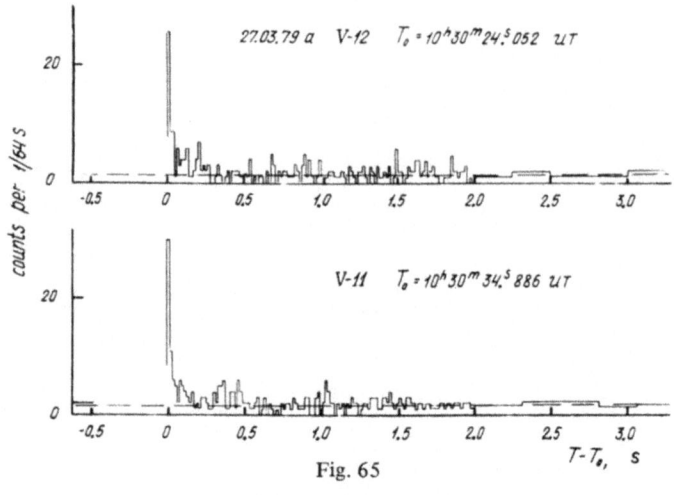

Fig. 65

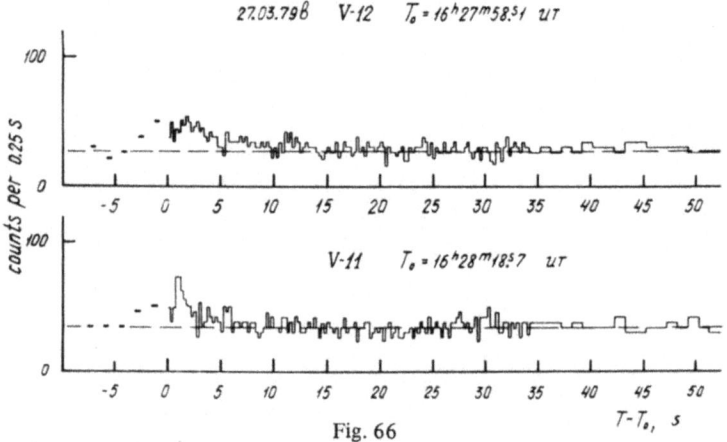

Fig. 66

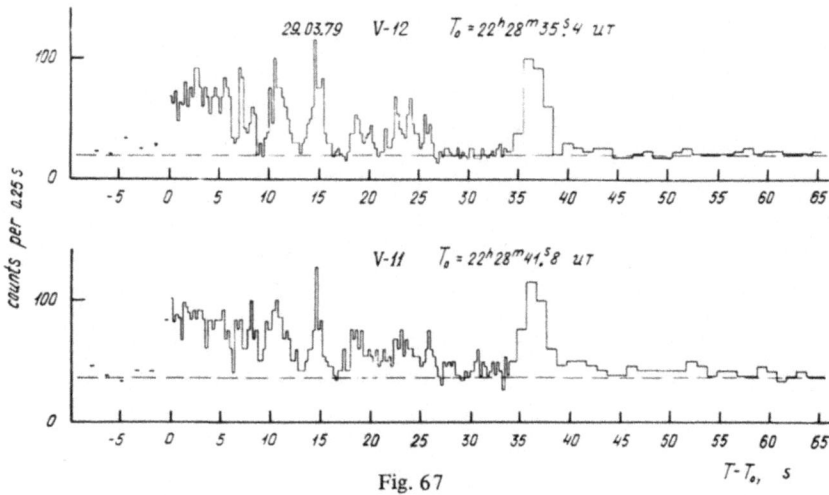

Fig. 67

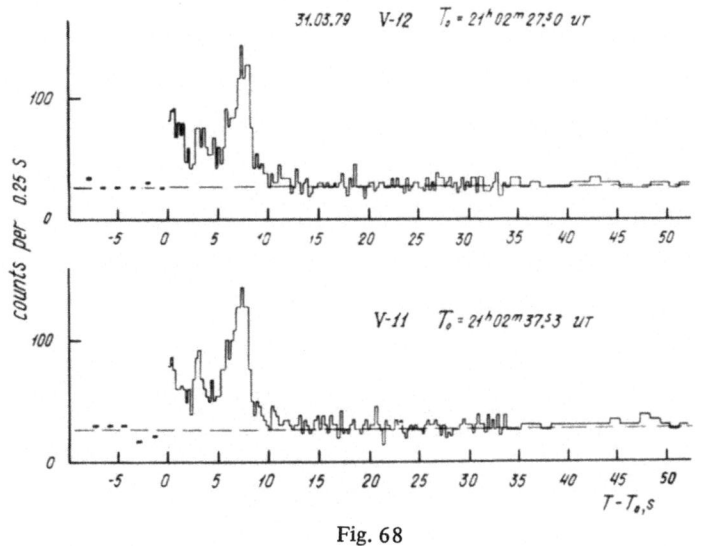

Fig. 68

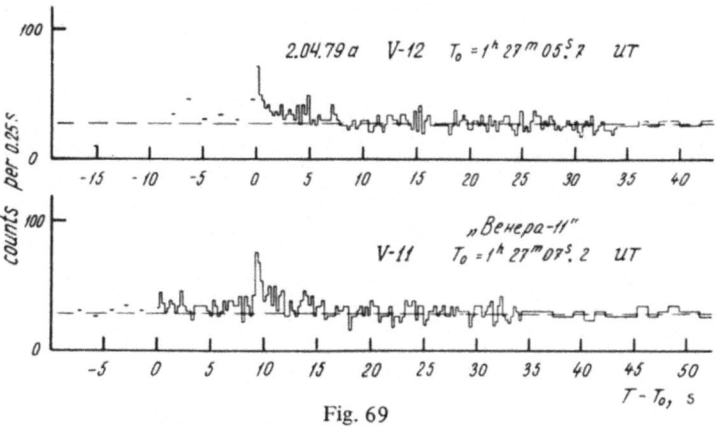

Fig. 69

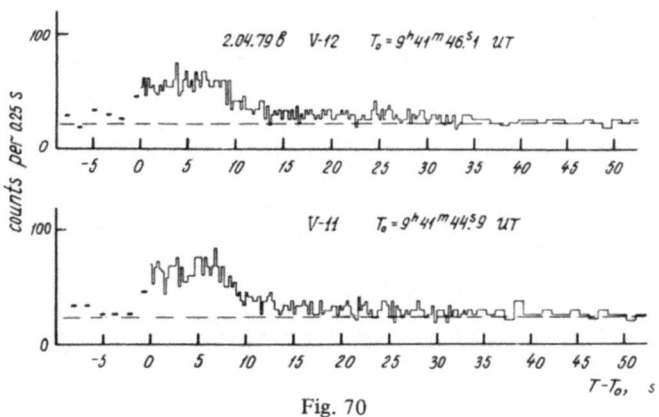

Fig. 70

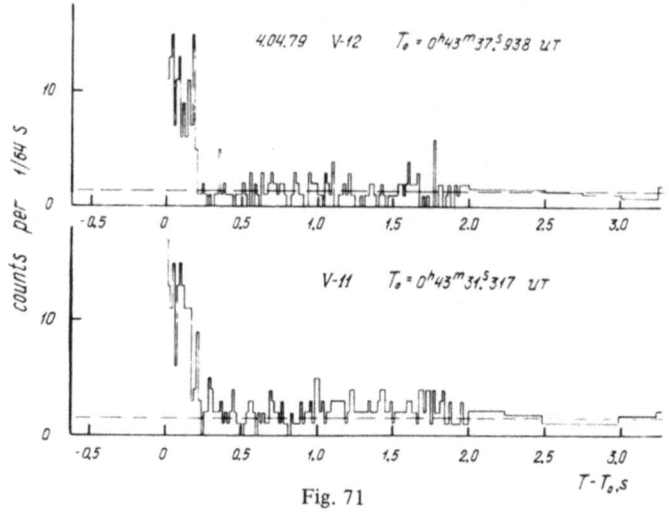

Fig. 71

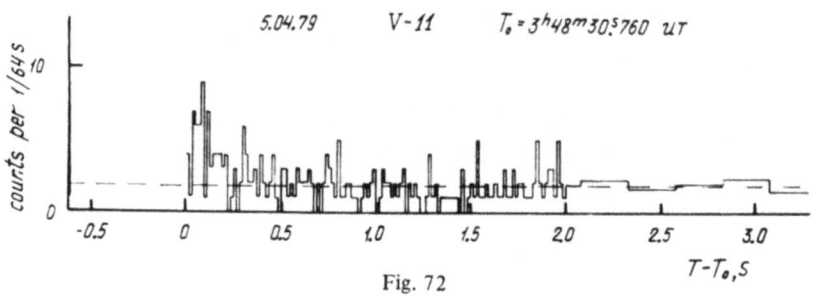

Fig. 72

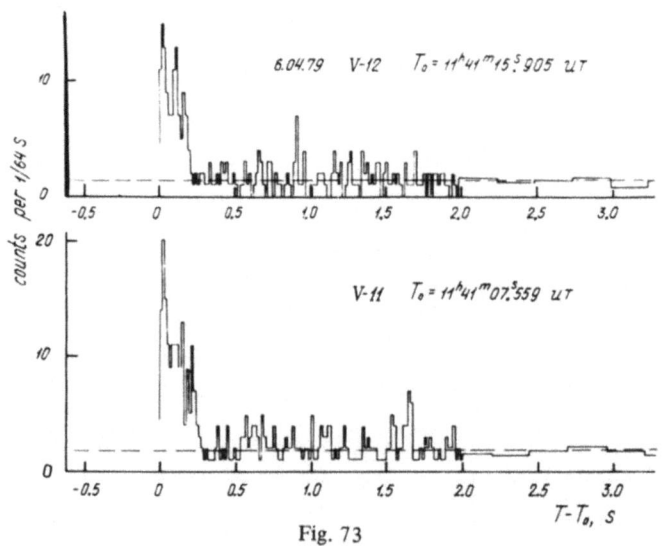

Fig. 73

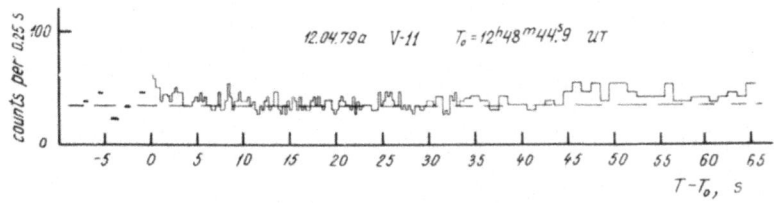

Fig. 74

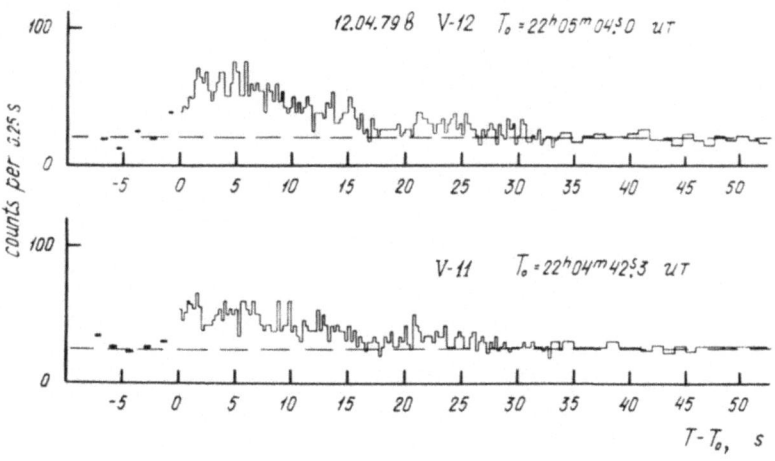

Fig. 75

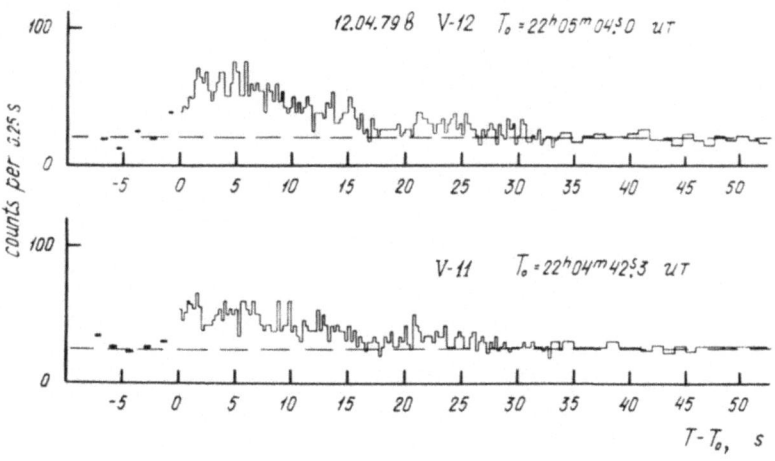

Fig. 76

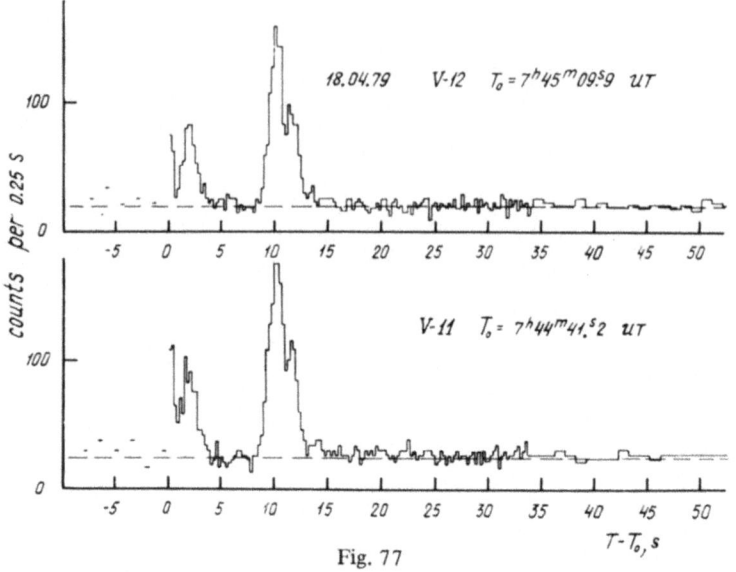

Fig. 77

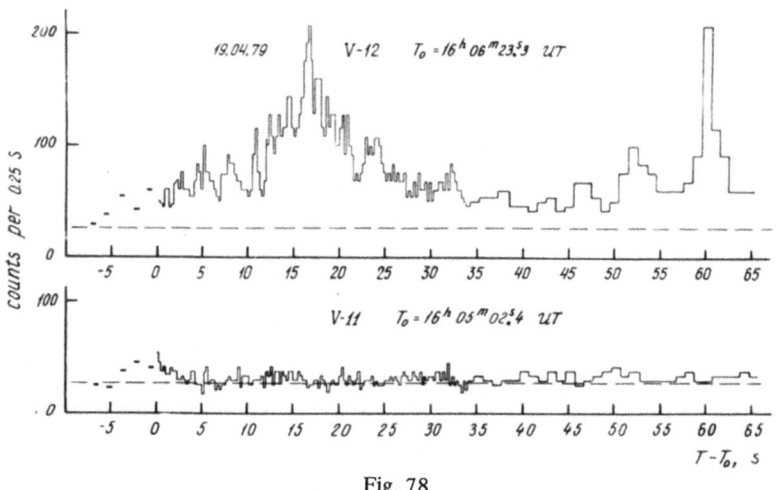

Fig. 78

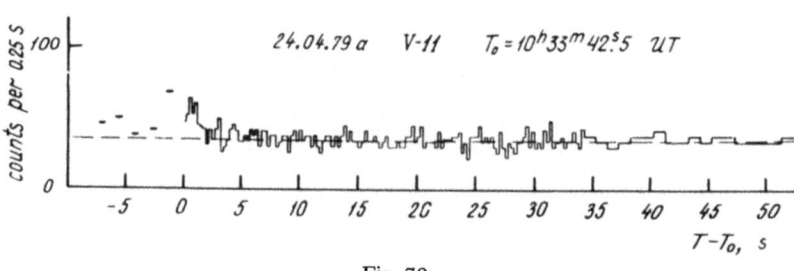

Fig. 79

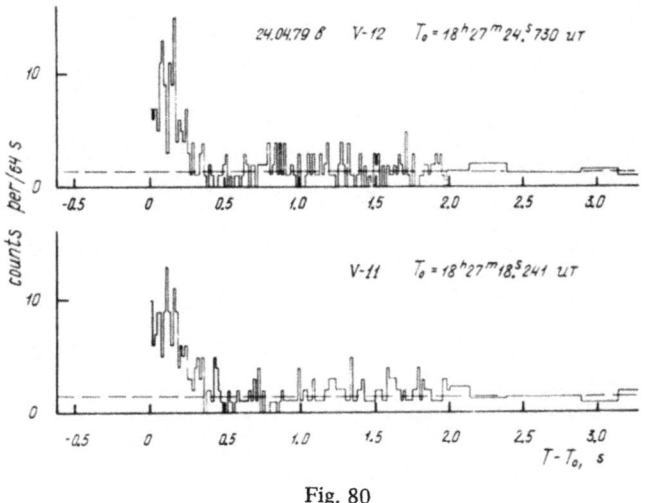

Fig. 80

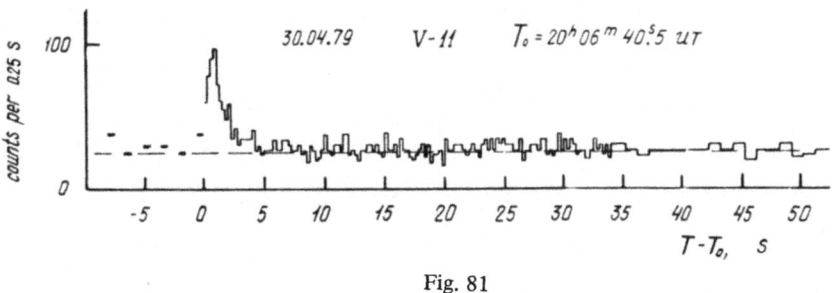

Fig. 81

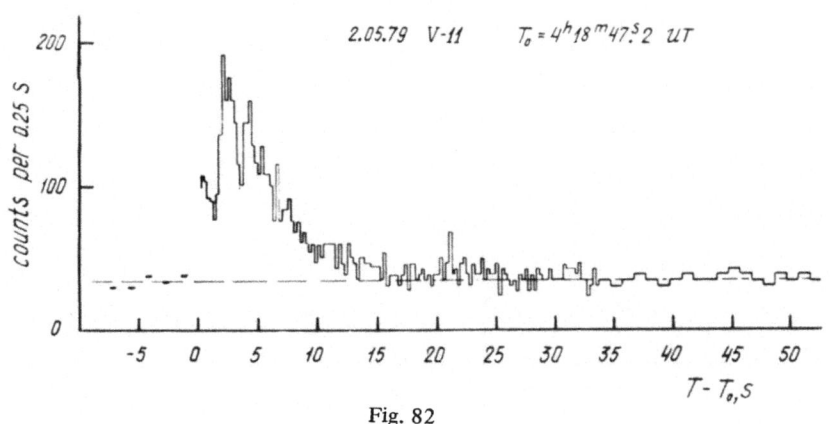

Fig. 82

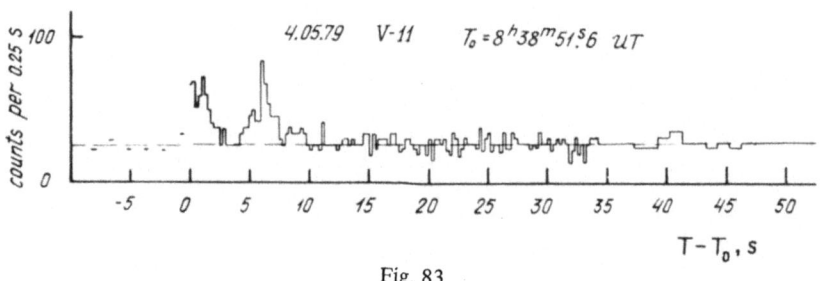

Fig. 83

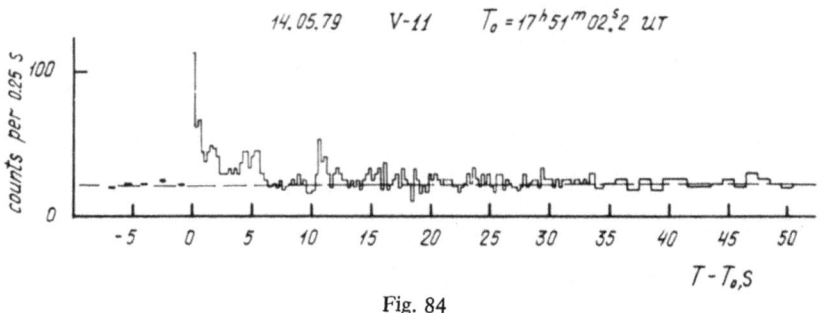

Fig. 84

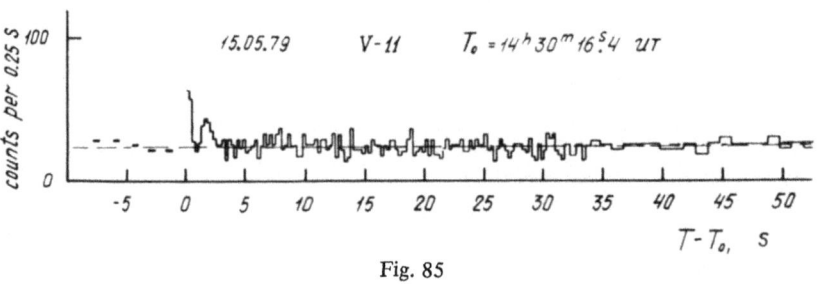

Fig. 85

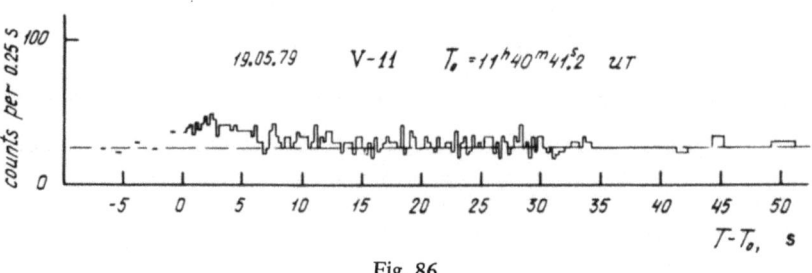

Fig. 86

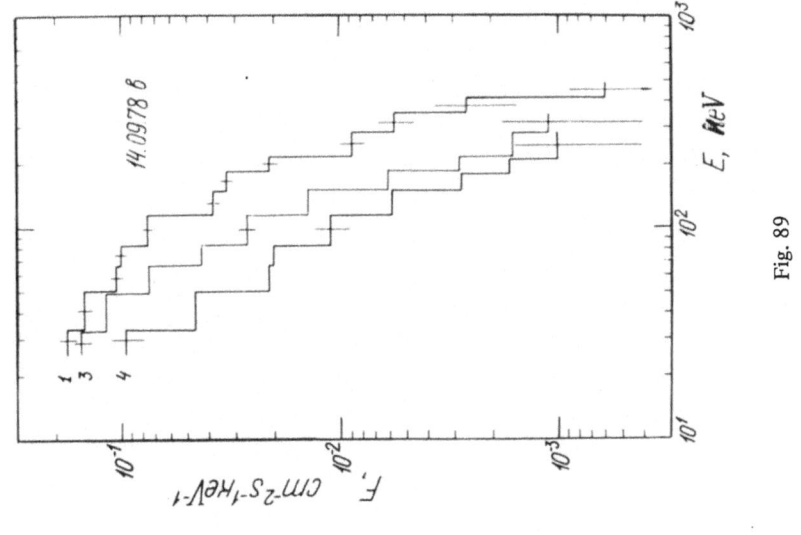

Fig. 89

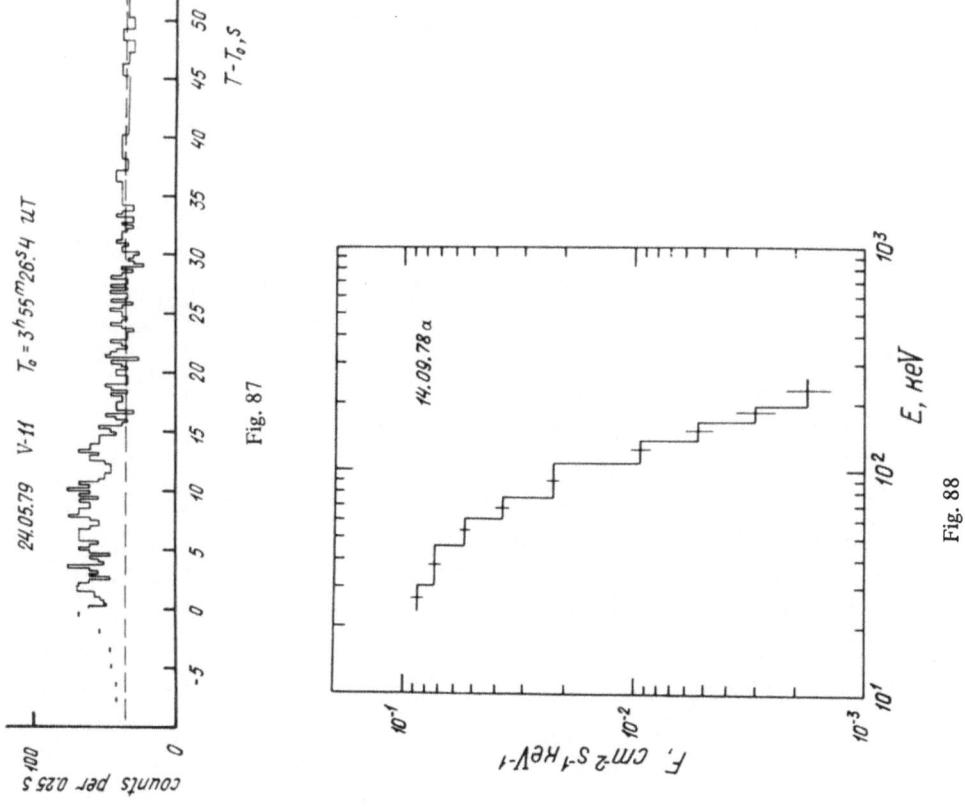

Fig. 87

Fig. 88

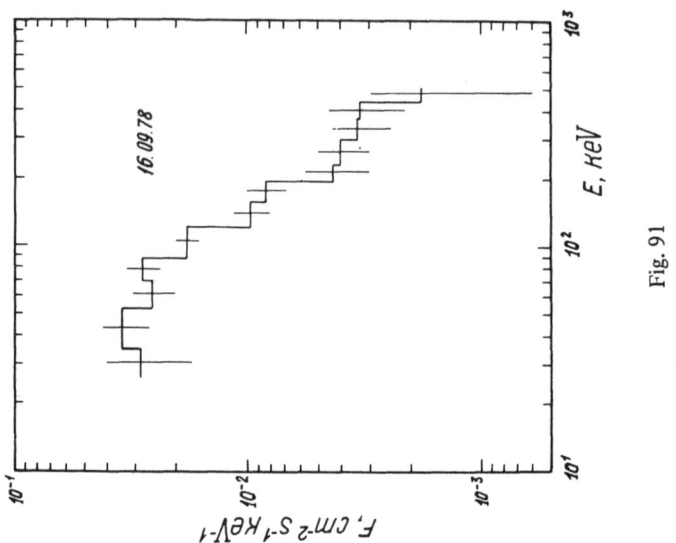

Fig. 91

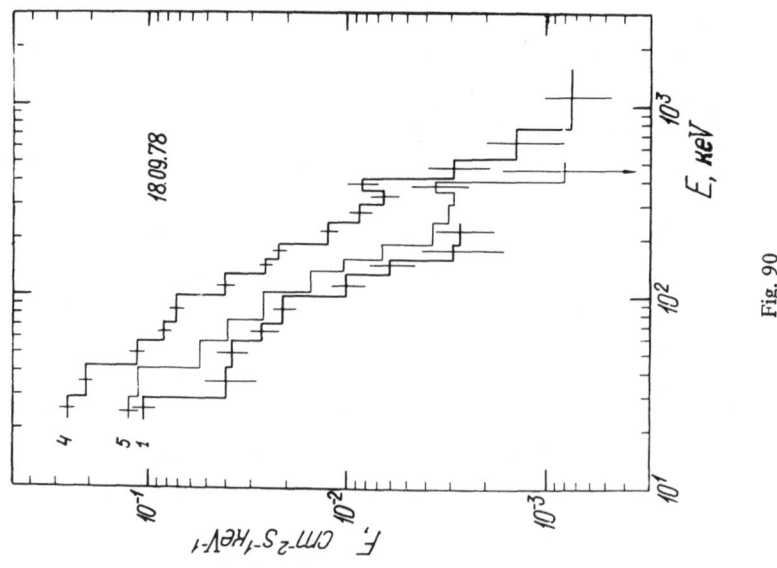

Fig. 90

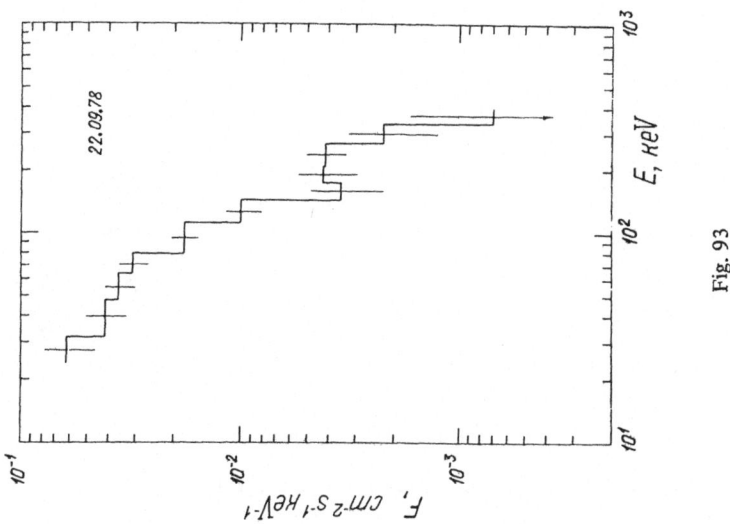

Fig. 93

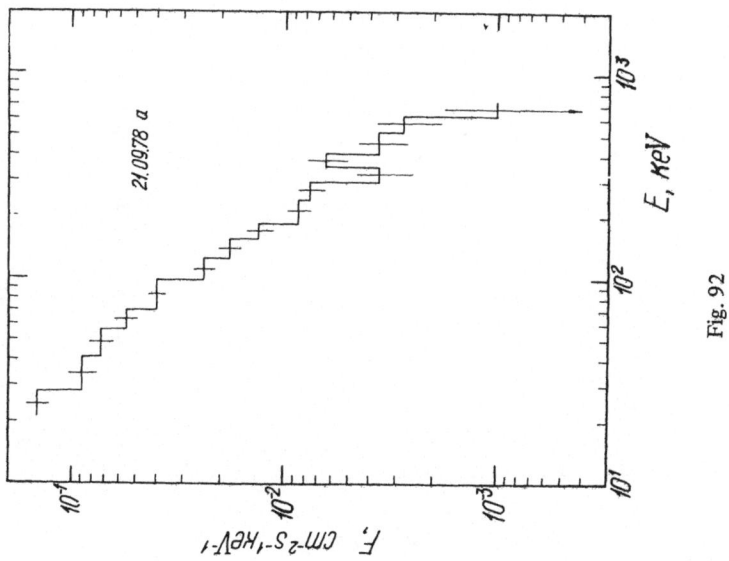

Fig. 92

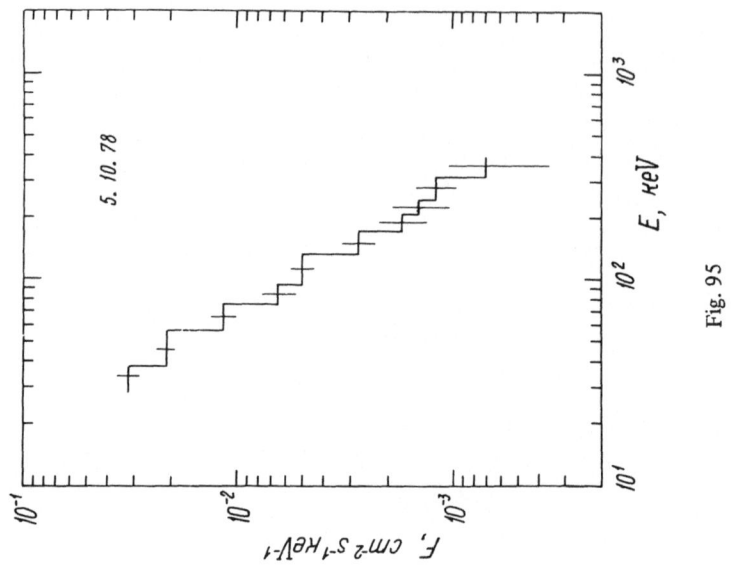

Fig. 95

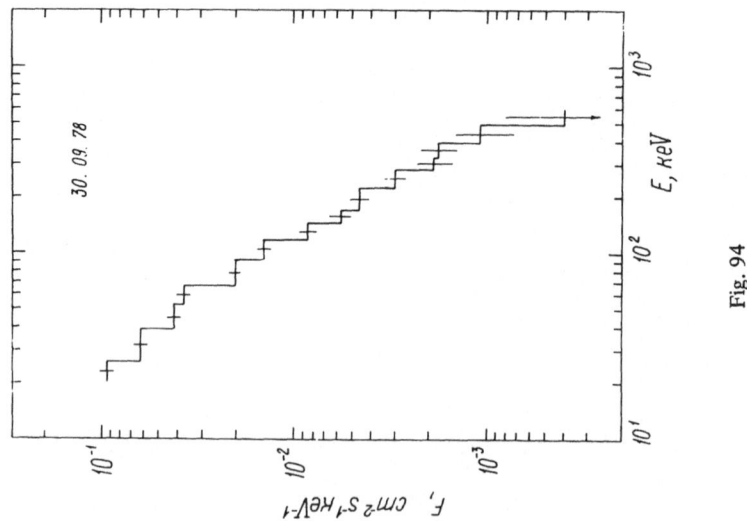

Fig. 94

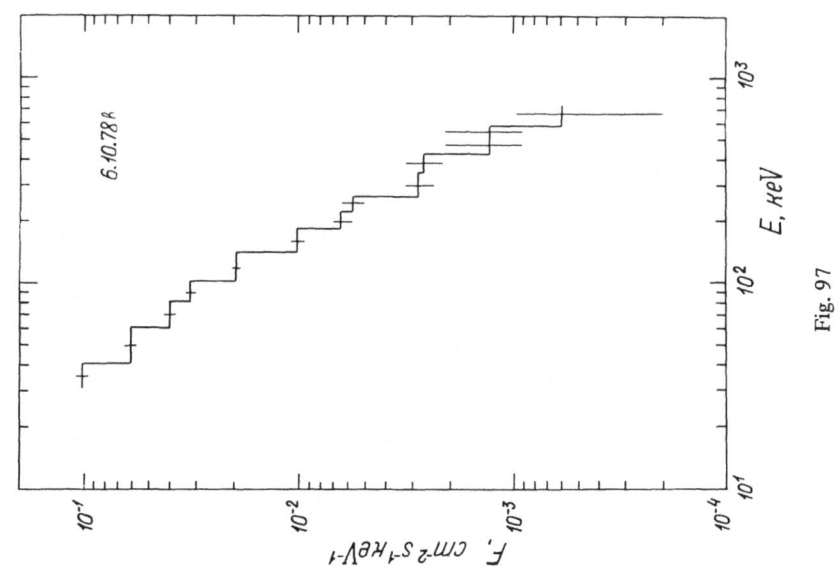

Fig. 97

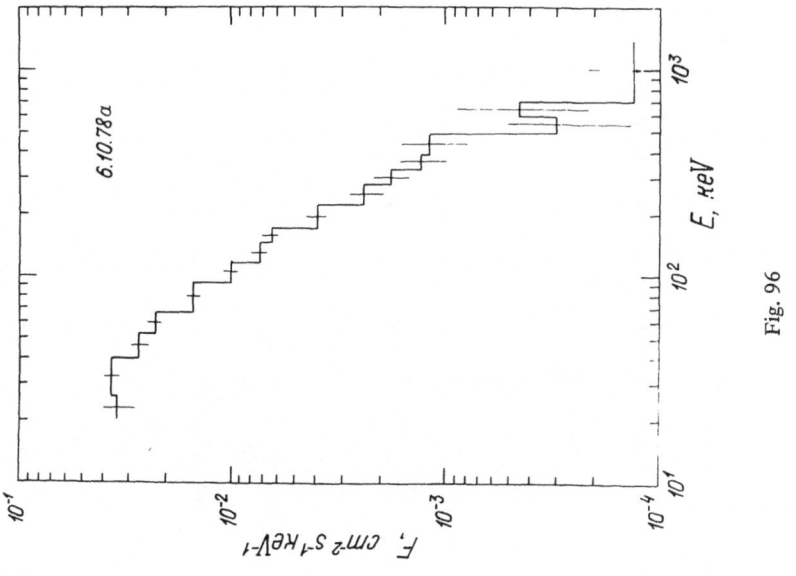

Fig. 96

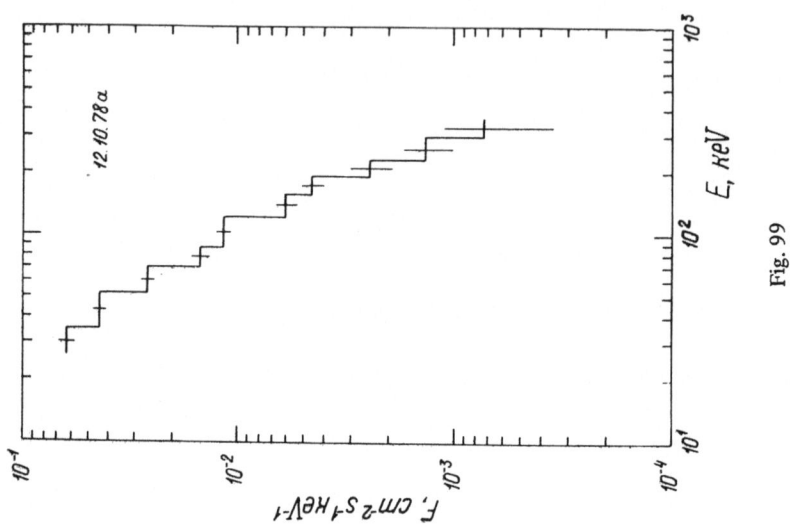

Fig. 99

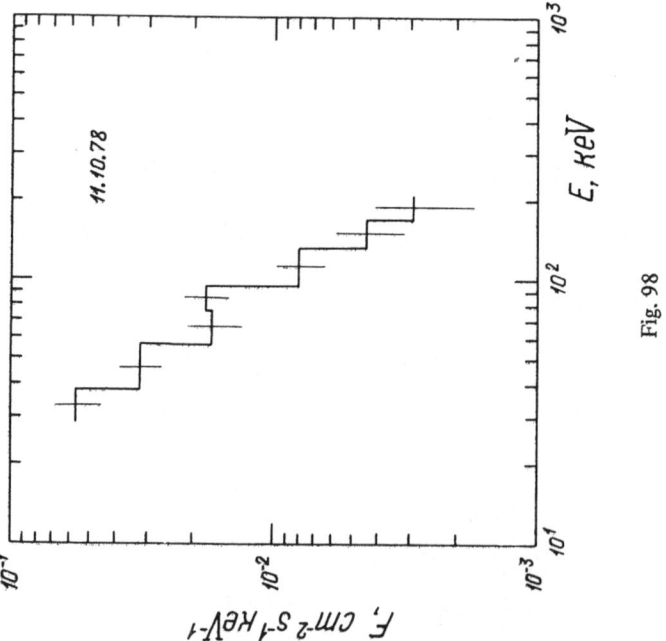

Fig. 98

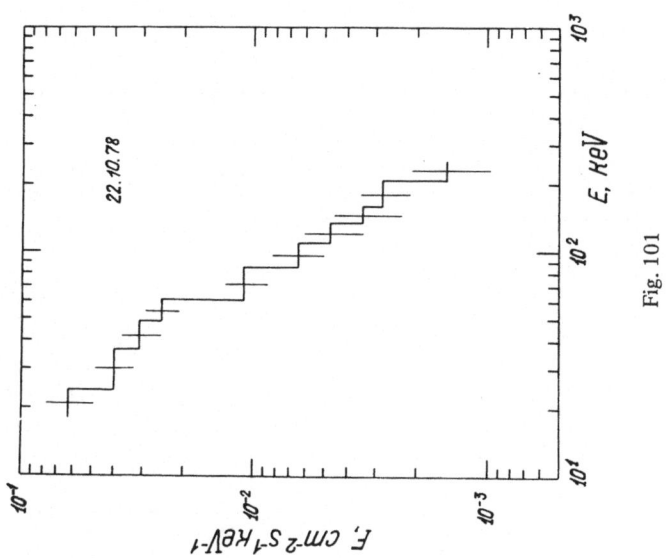

Fig. 101

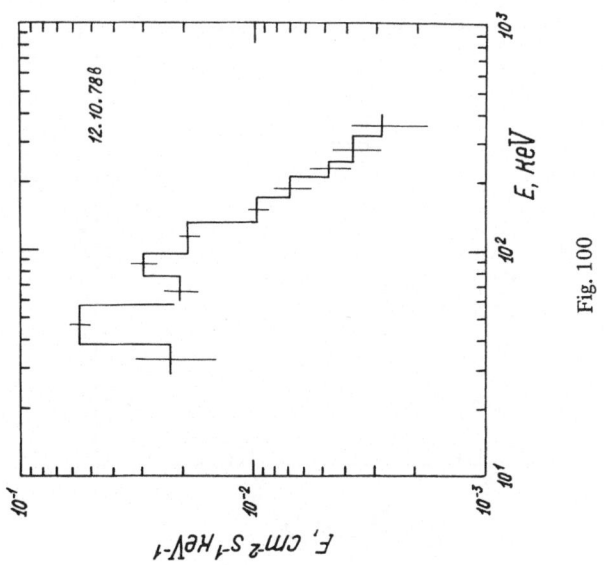

Fig. 100

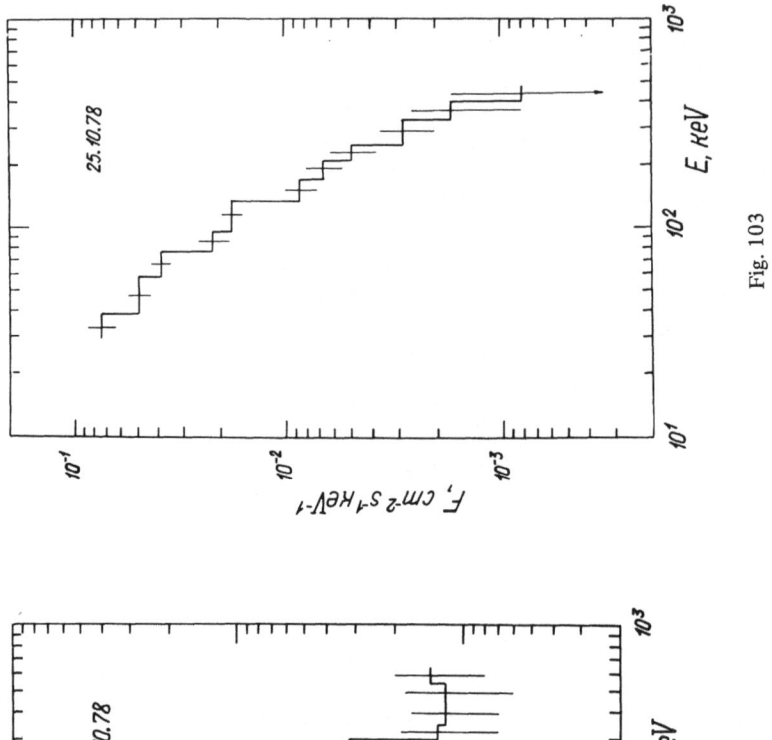

Fig. 103

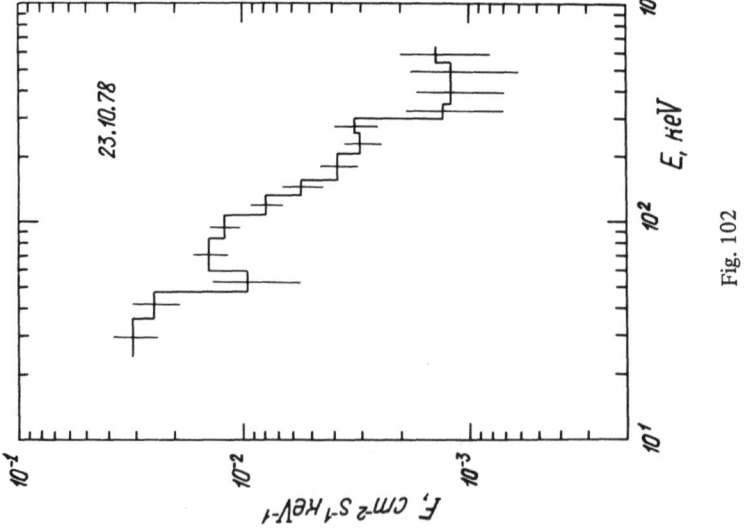

Fig. 102

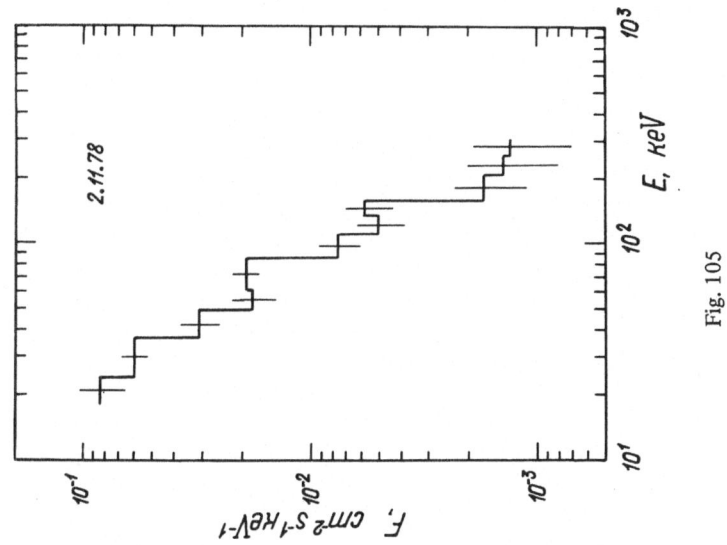

Fig. 105

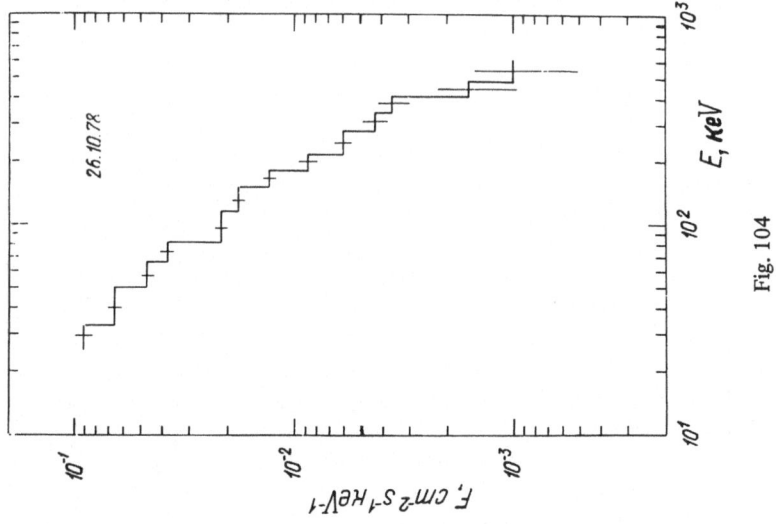

Fig. 104

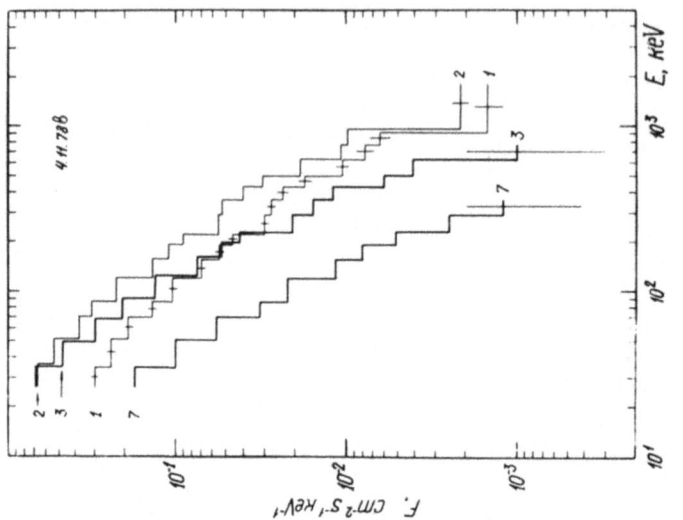

Fig. 107

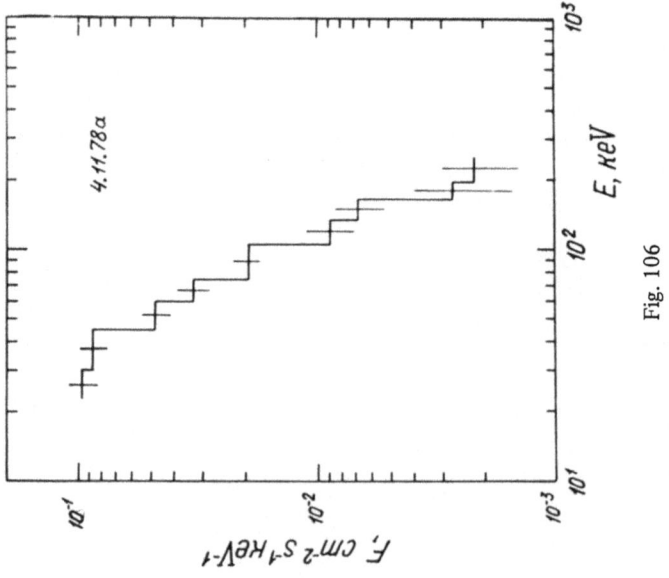

Fig. 106

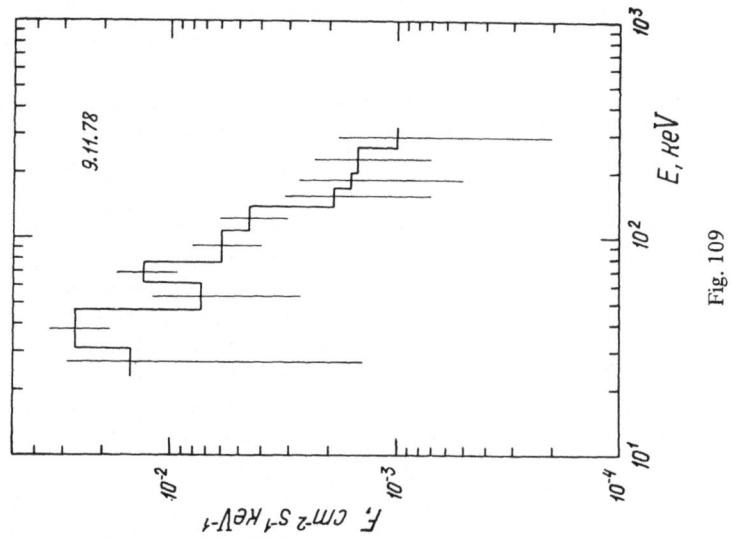

Fig. 109

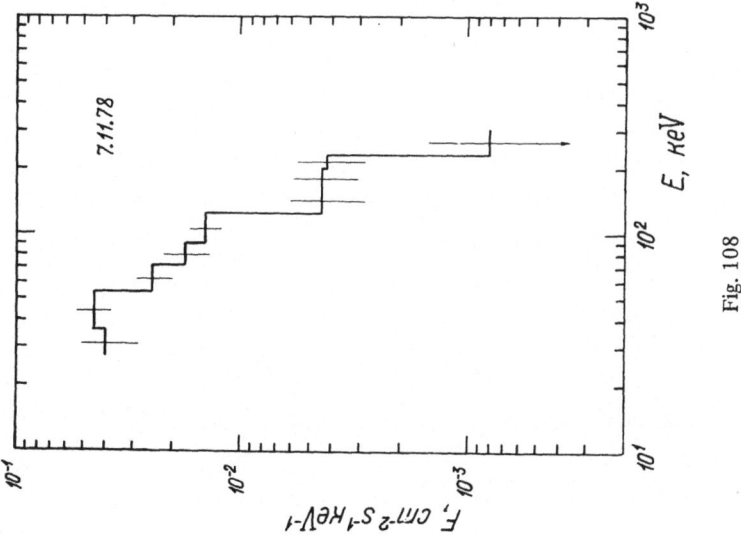

Fig. 108

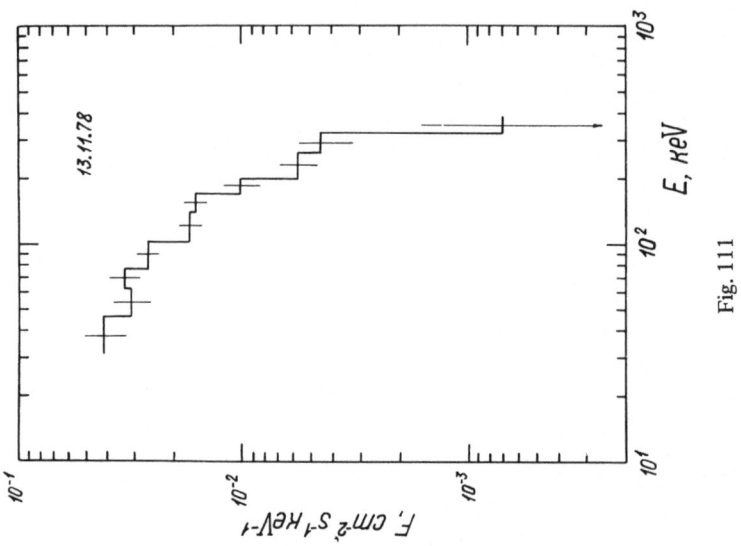

Fig. 111

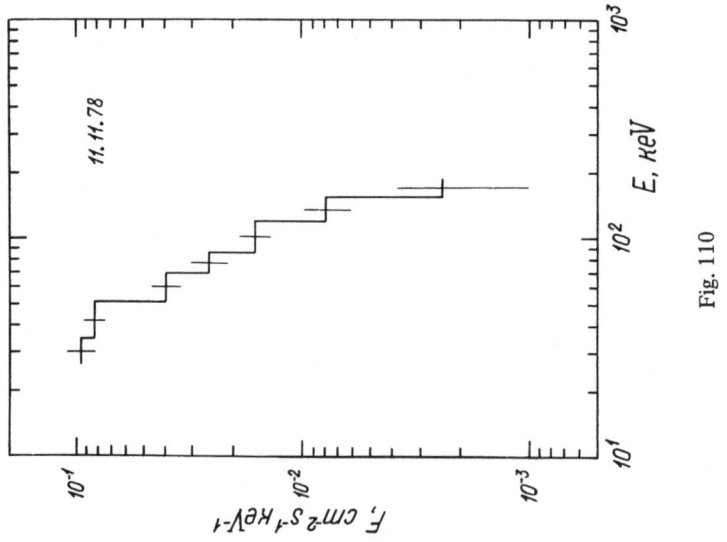

Fig. 110

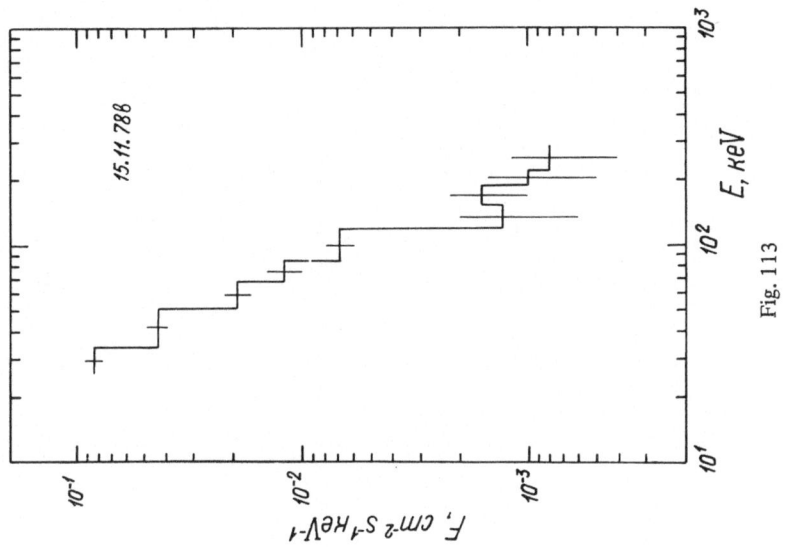

Fig. 113

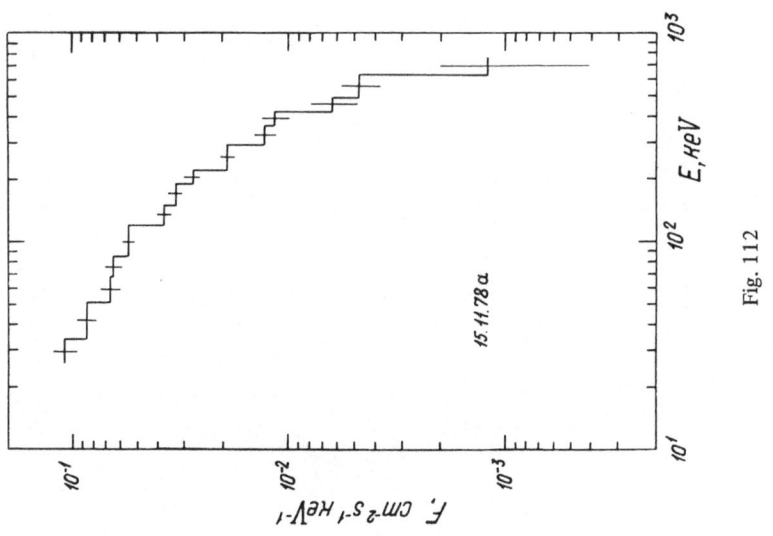

Fig. 112

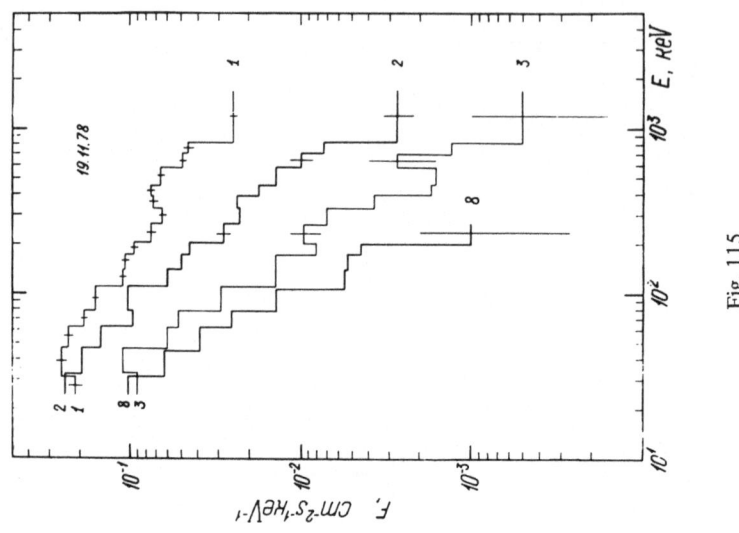

Fig. 115

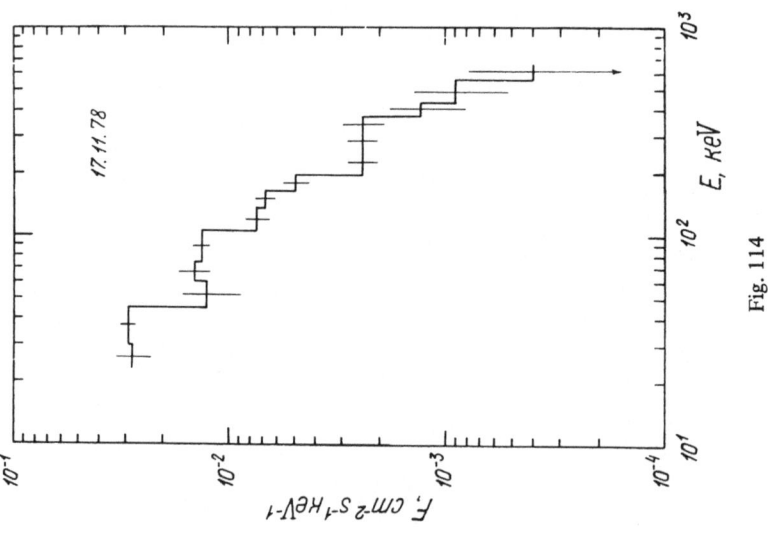

Fig. 114

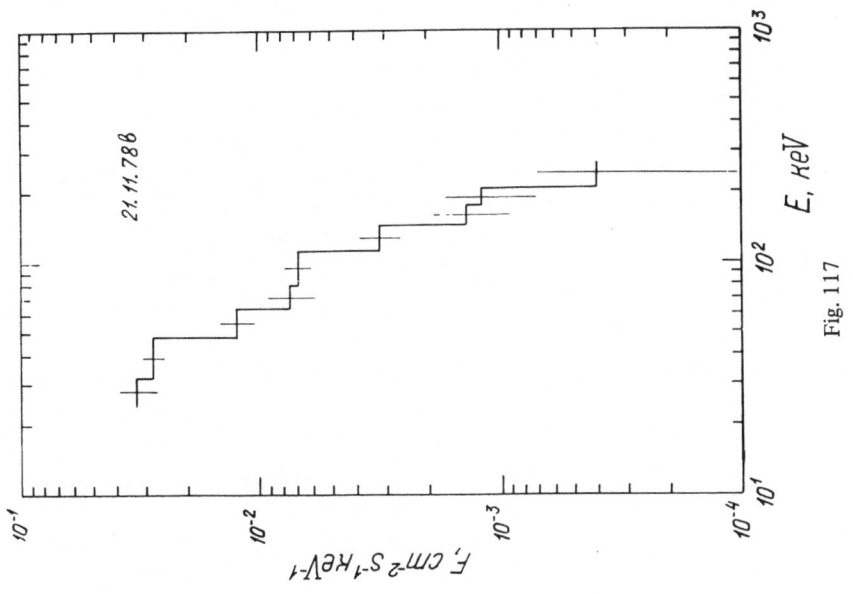

Fig. 117

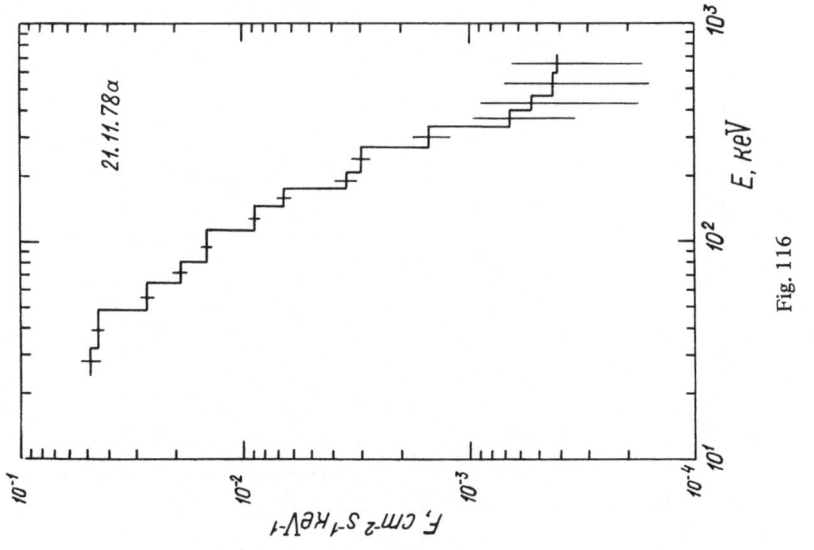

Fig. 116

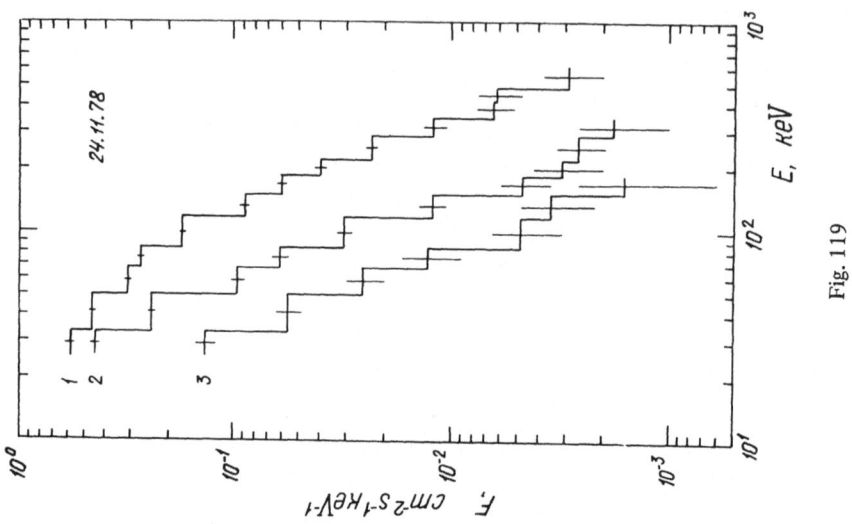

Fig. 119

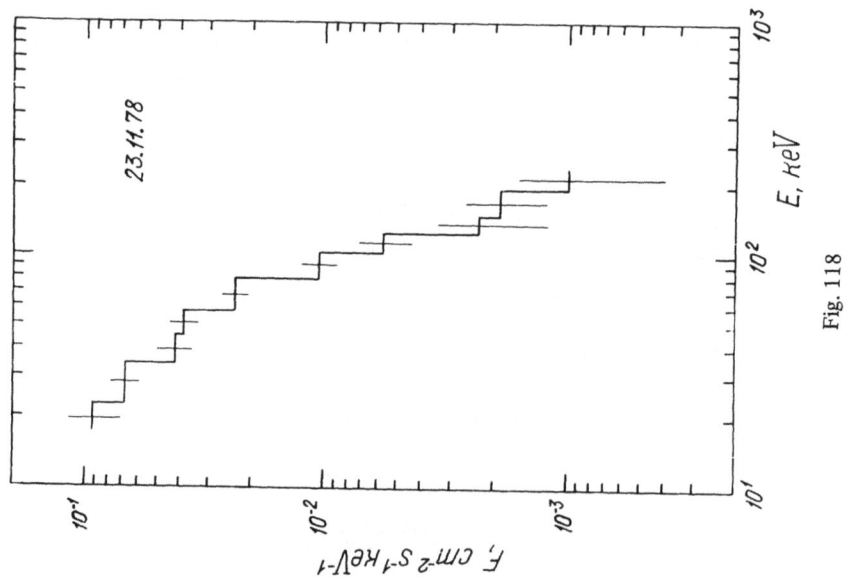

Fig. 118

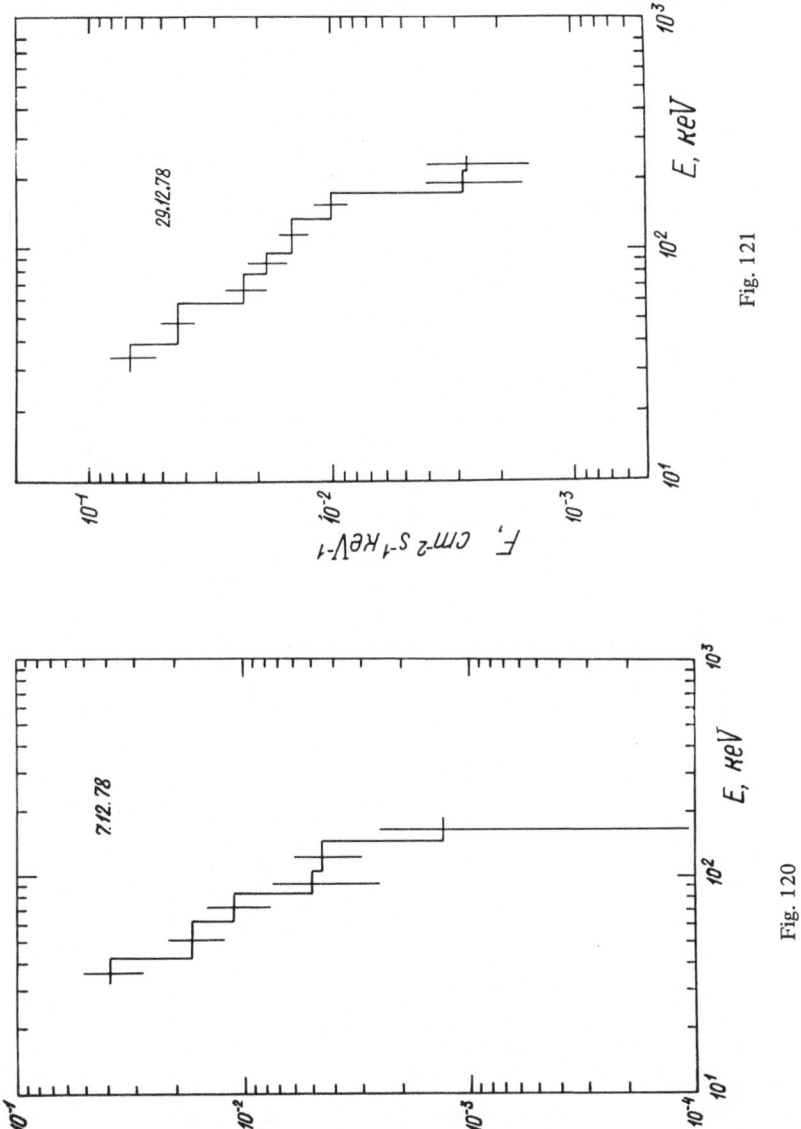

Fig. 121

Fig. 120

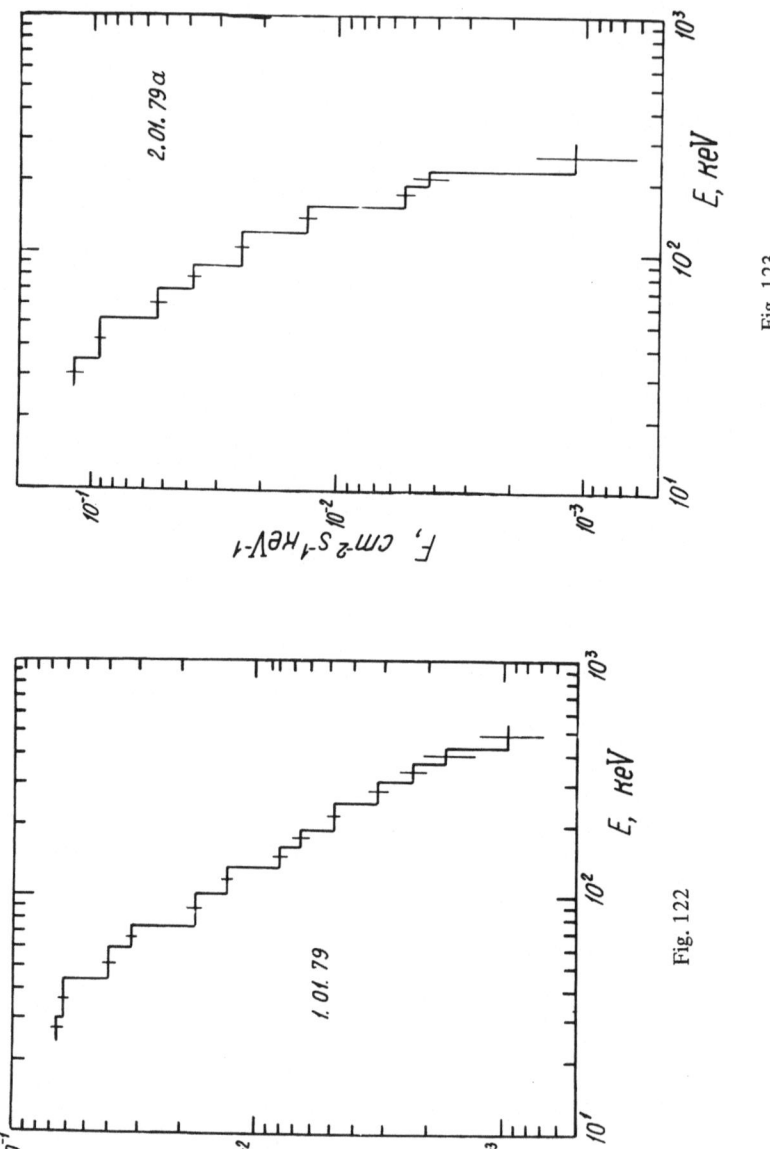

Fig. 123

Fig. 122

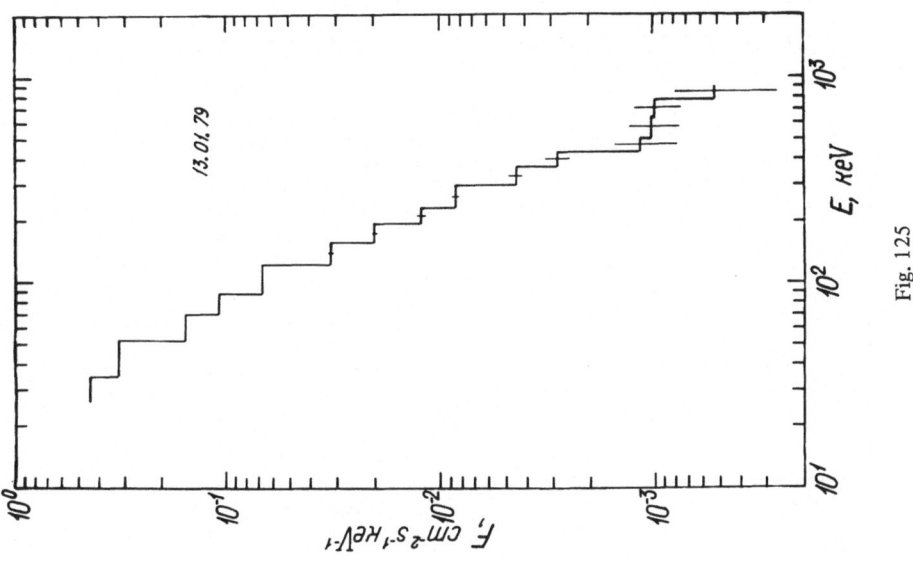

Fig. 125

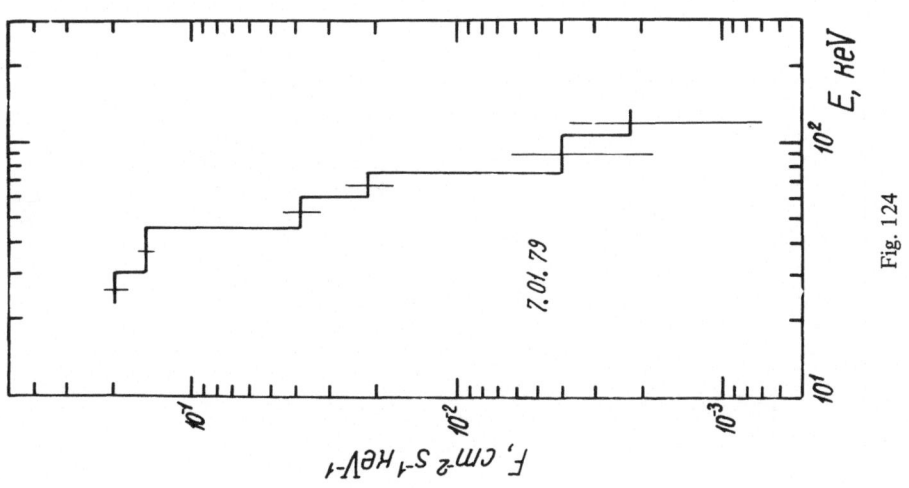

Fig. 124

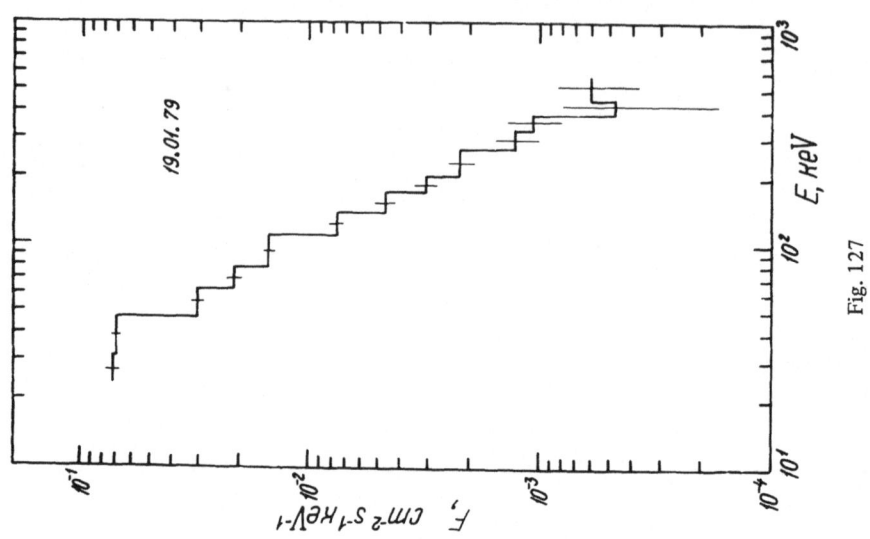

Fig. 127

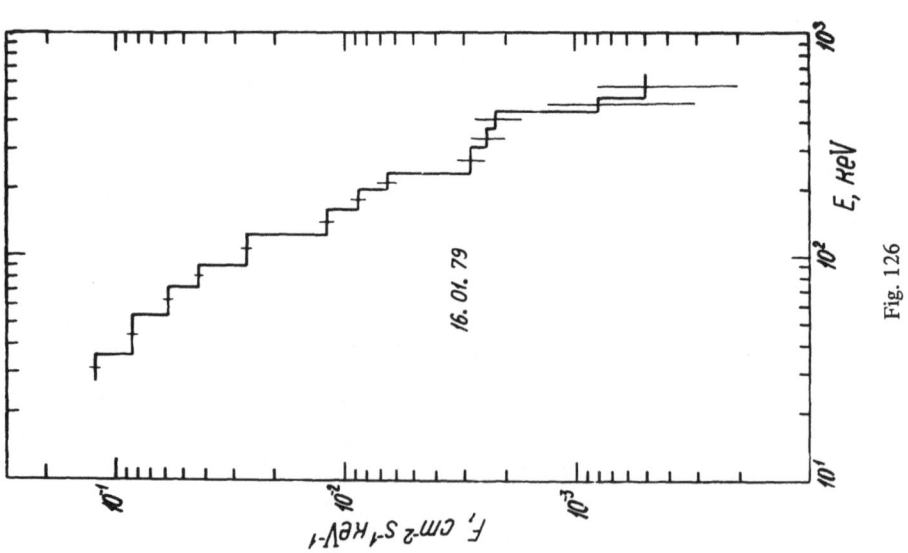

Fig. 126

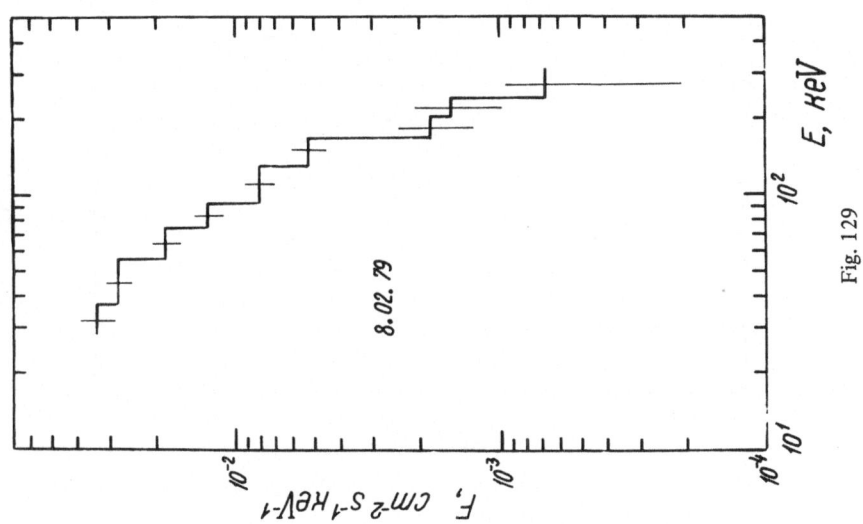

Fig. 129

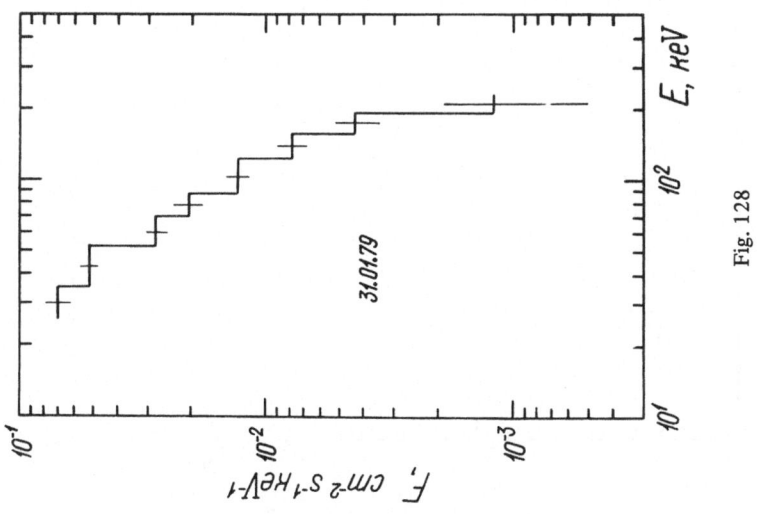

Fig. 128

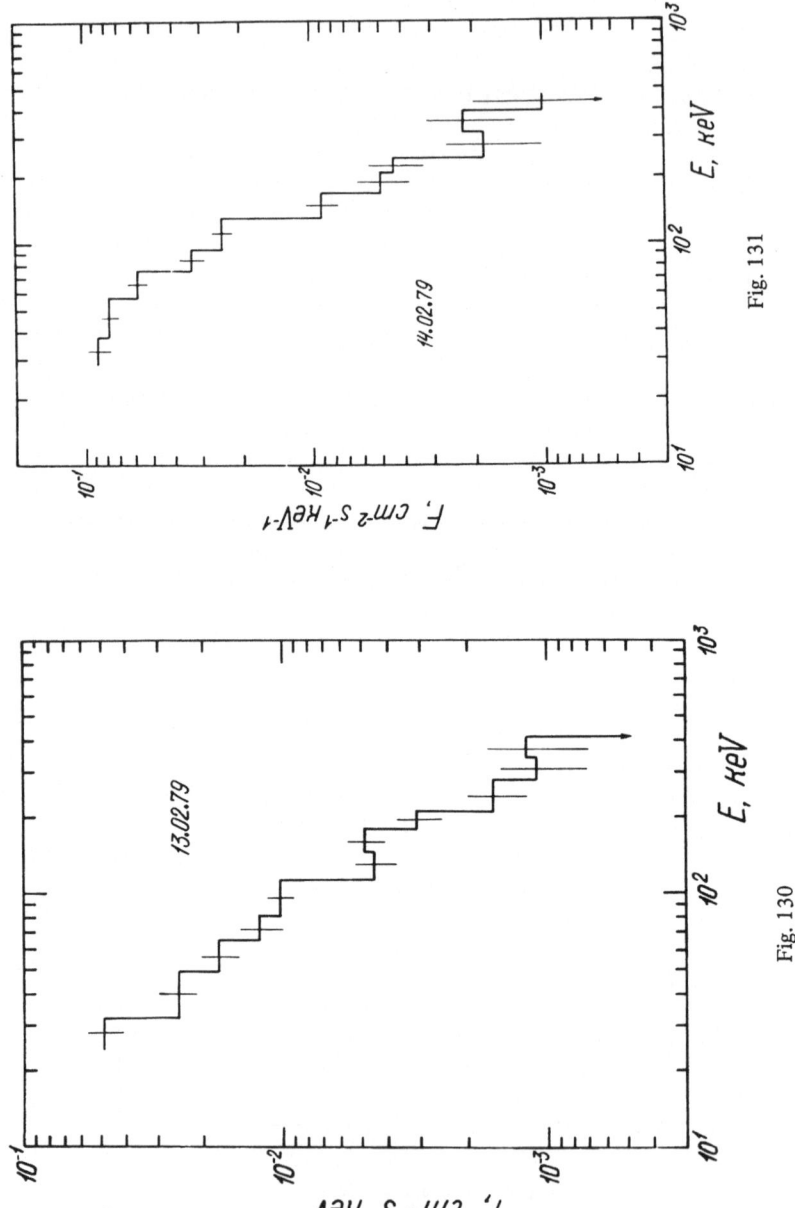

Fig. 130

Fig. 131

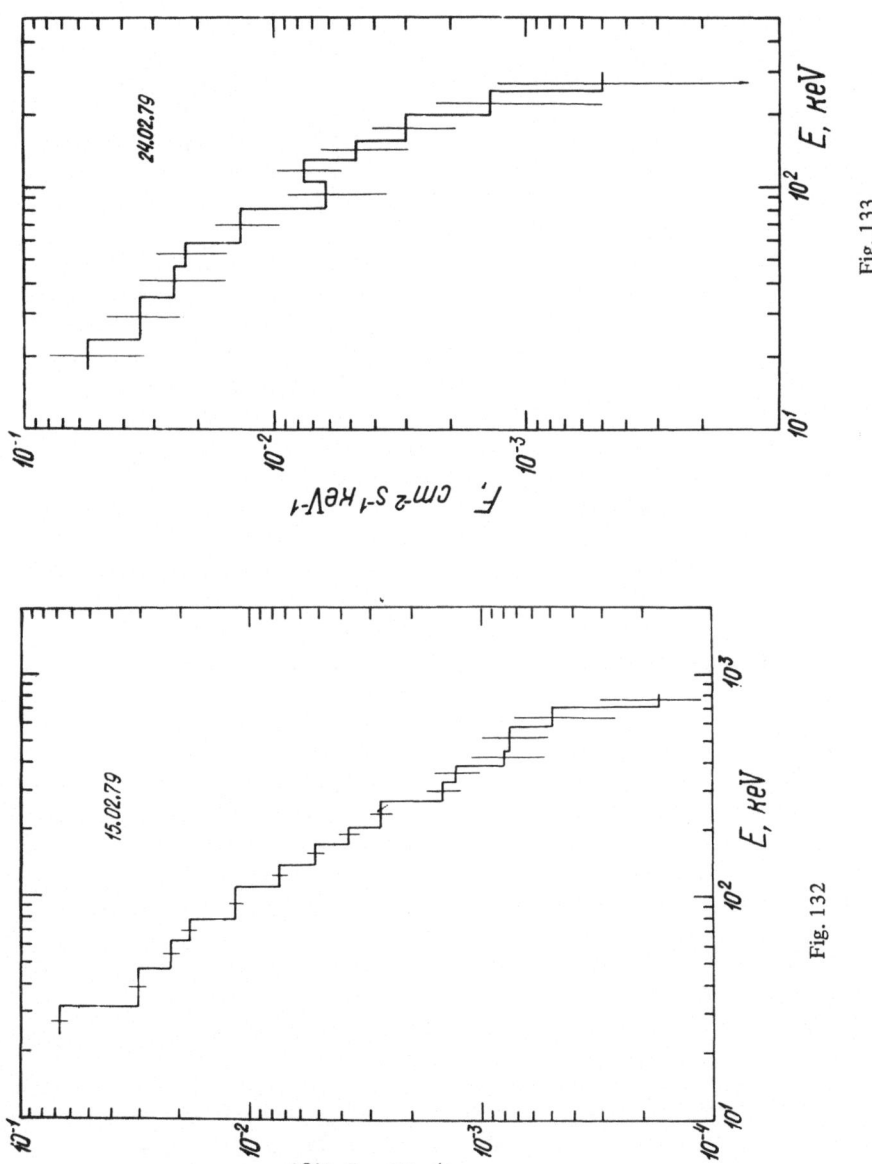

Fig. 133

Fig. 132

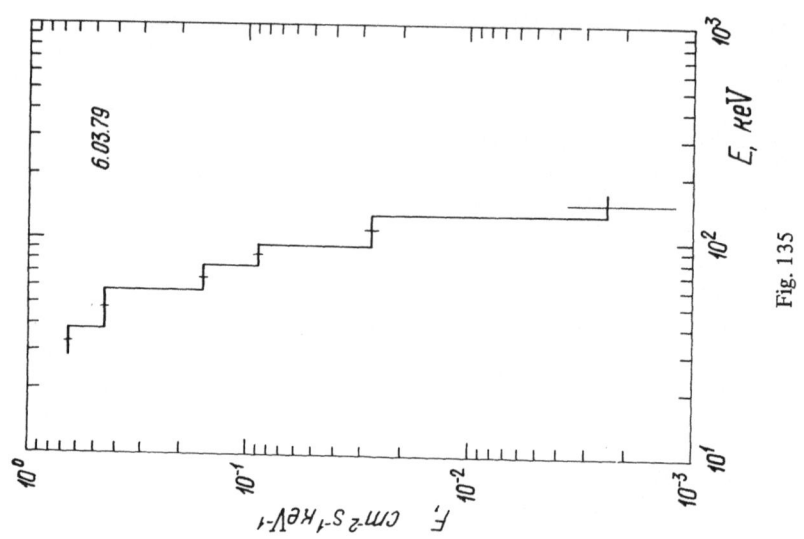

Fig. 135

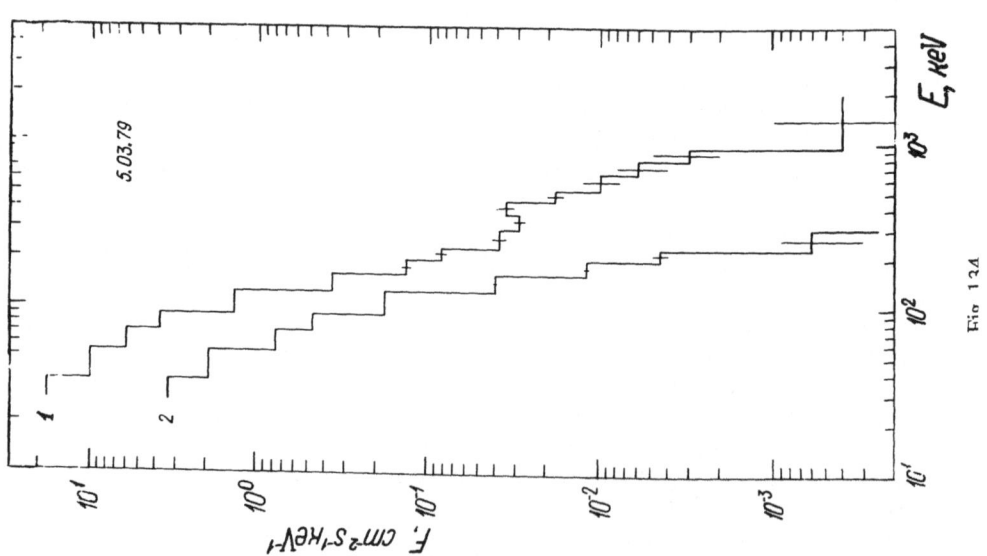

Fig. 134

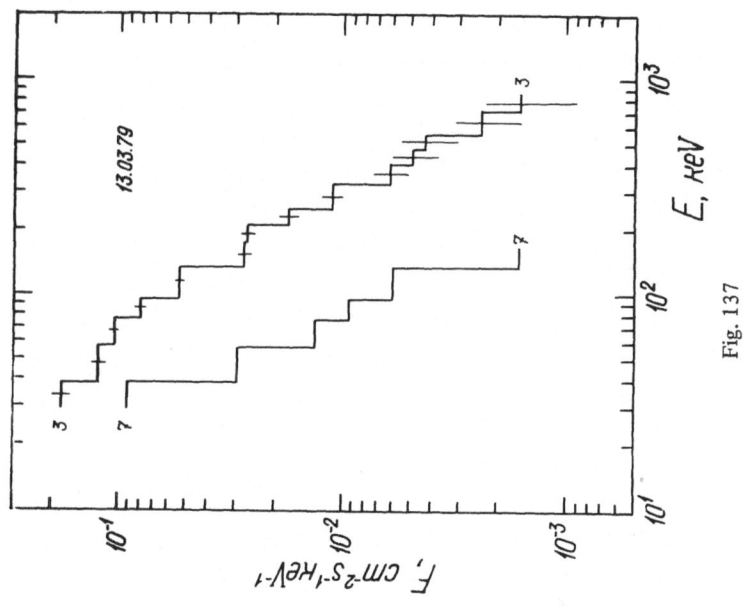

Fig. 137

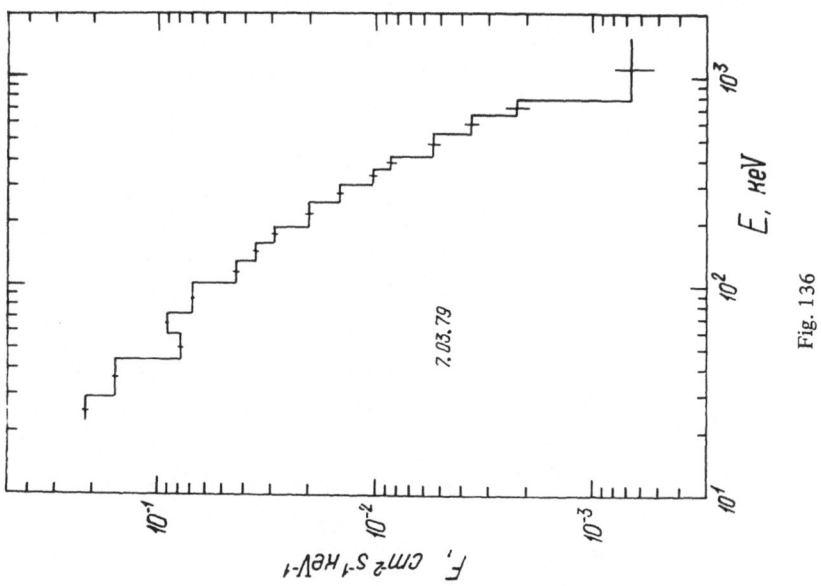

Fig. 136

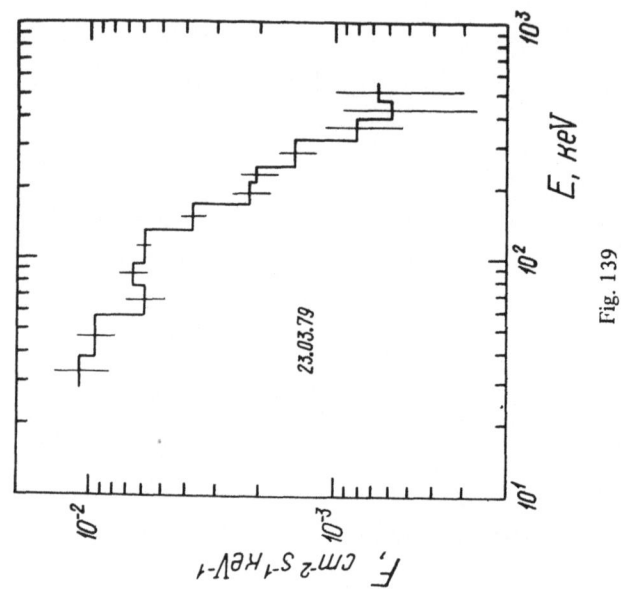

Fig. 139

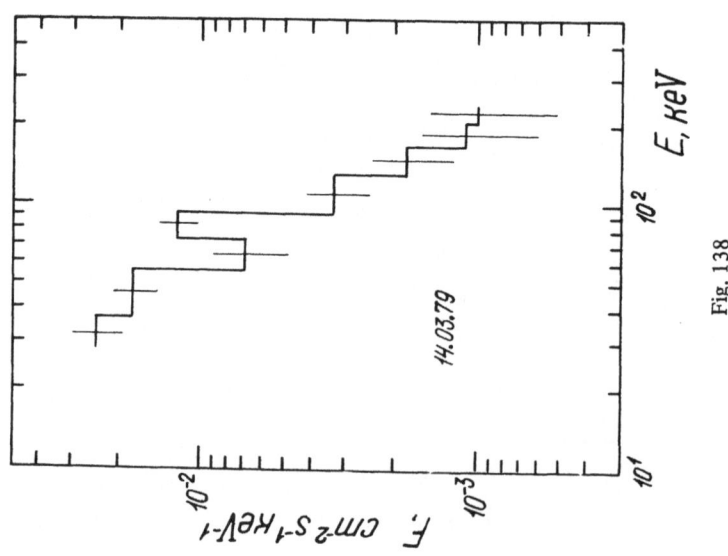

Fig. 138

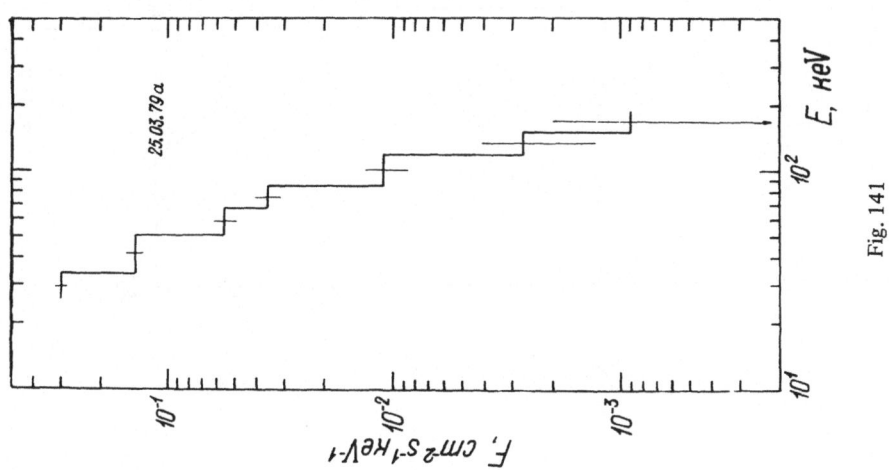

Fig. 141

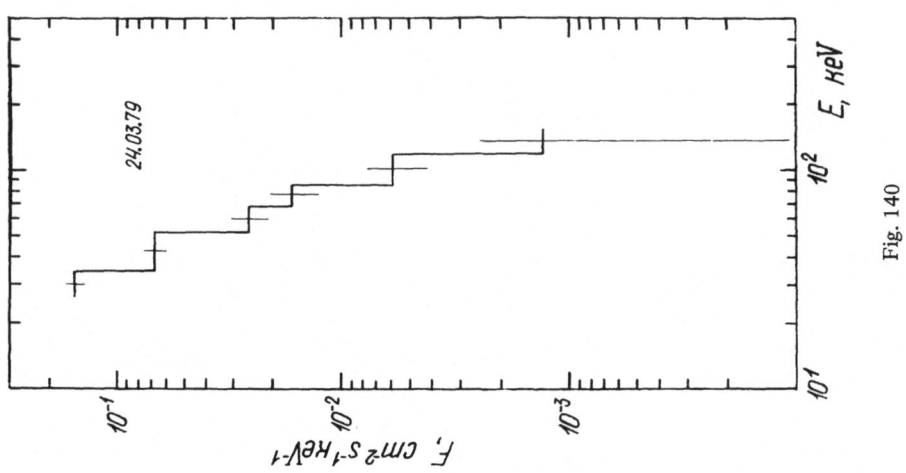

Fig. 140

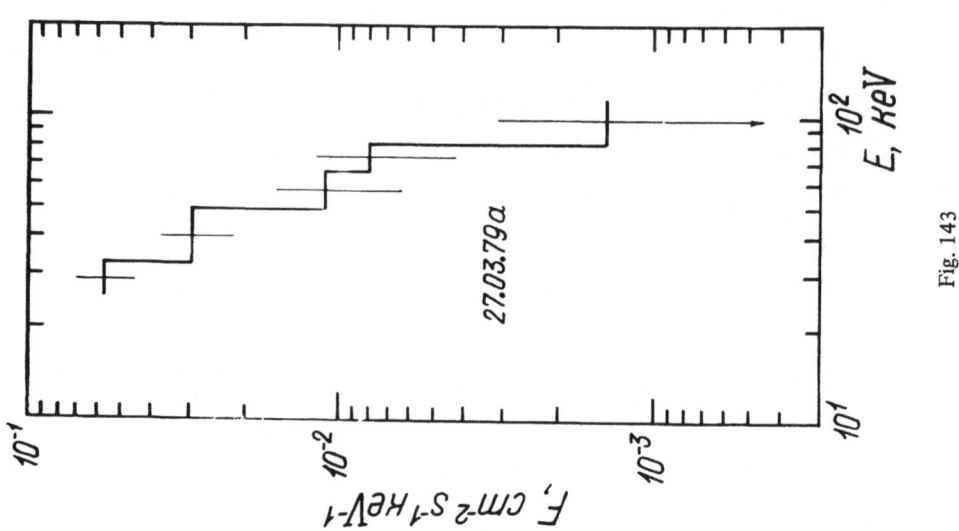

Fig. 143

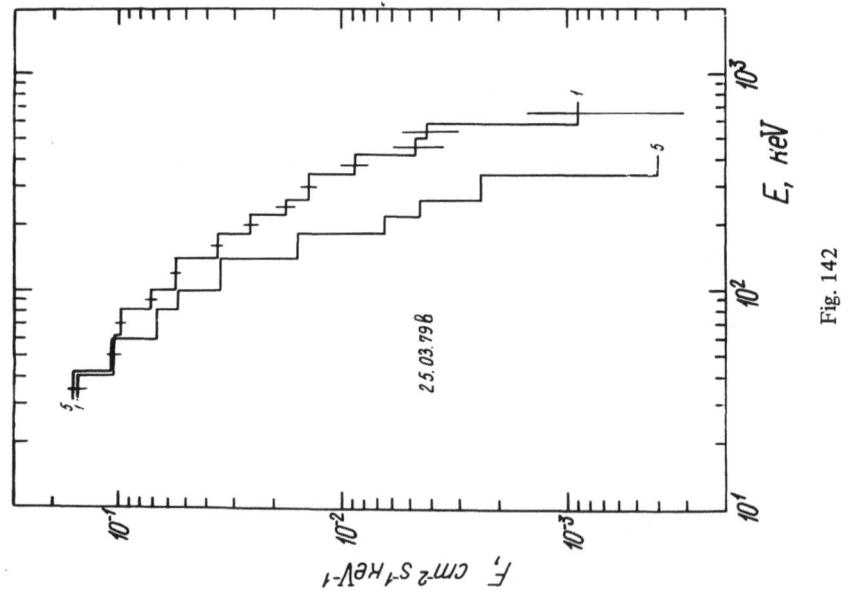

Fig. 142

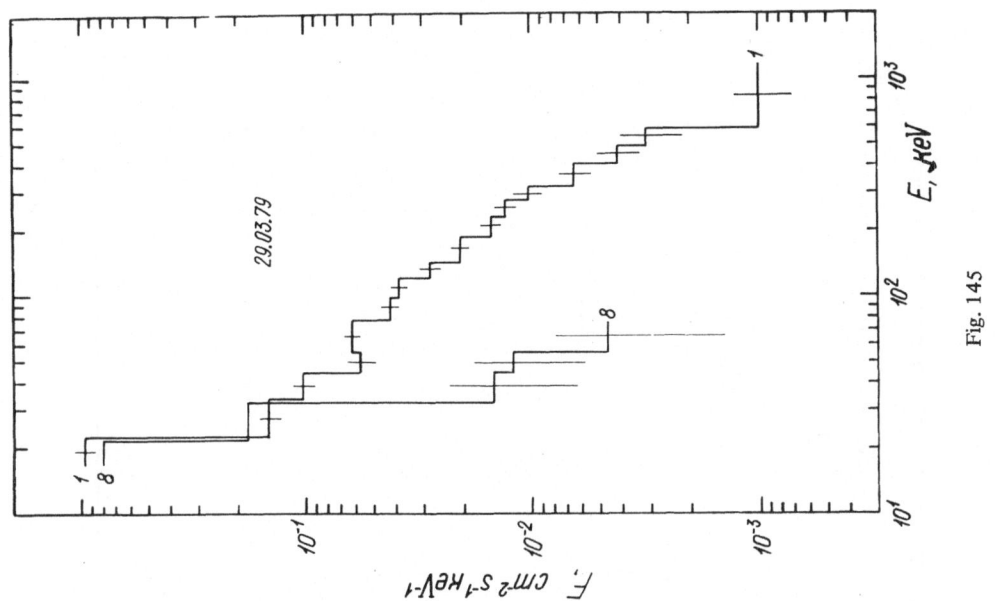

Fig. 145

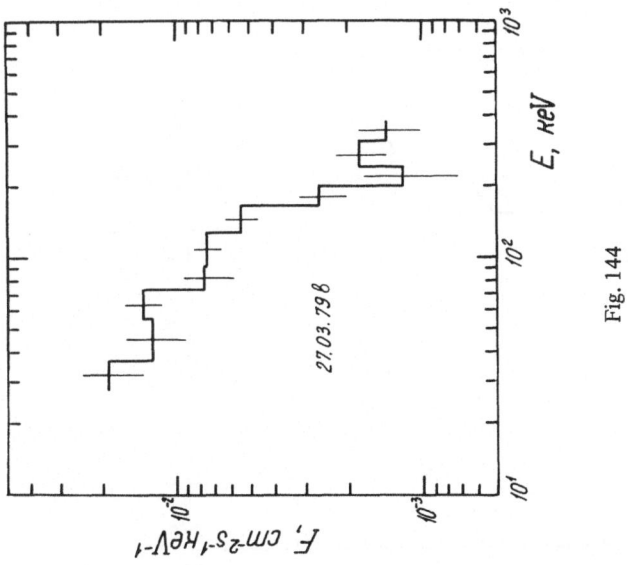

Fig. 144

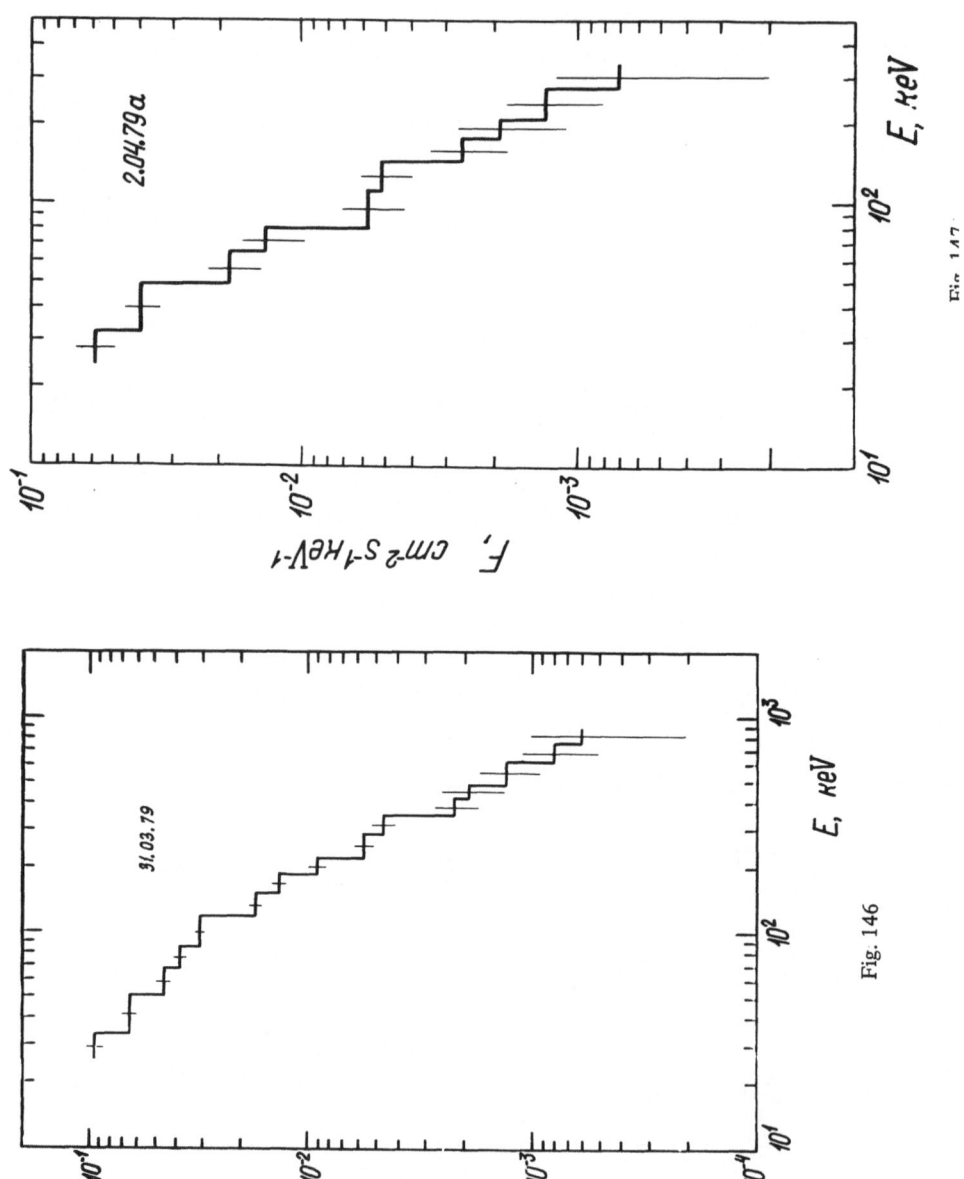

Fig. 146

Fig. 147

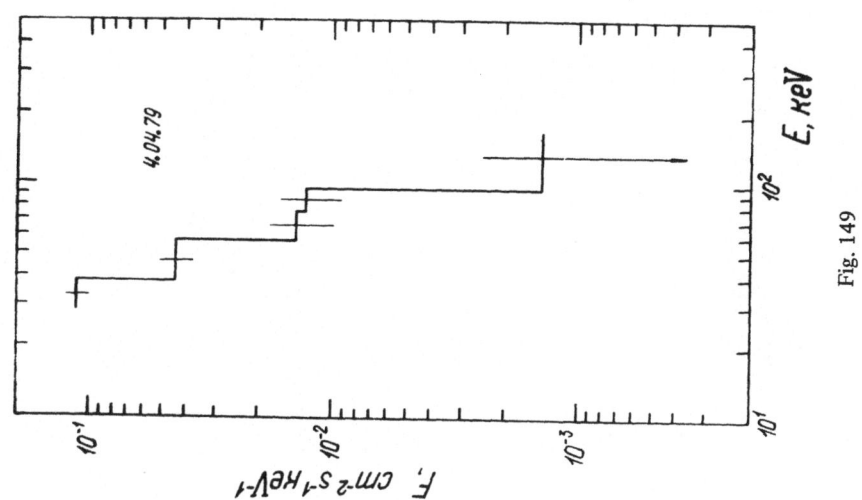

Fig. 149

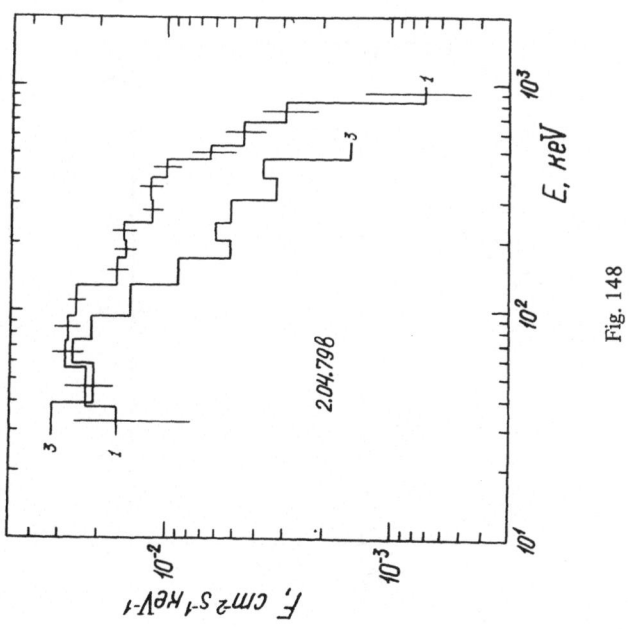

Fig. 148

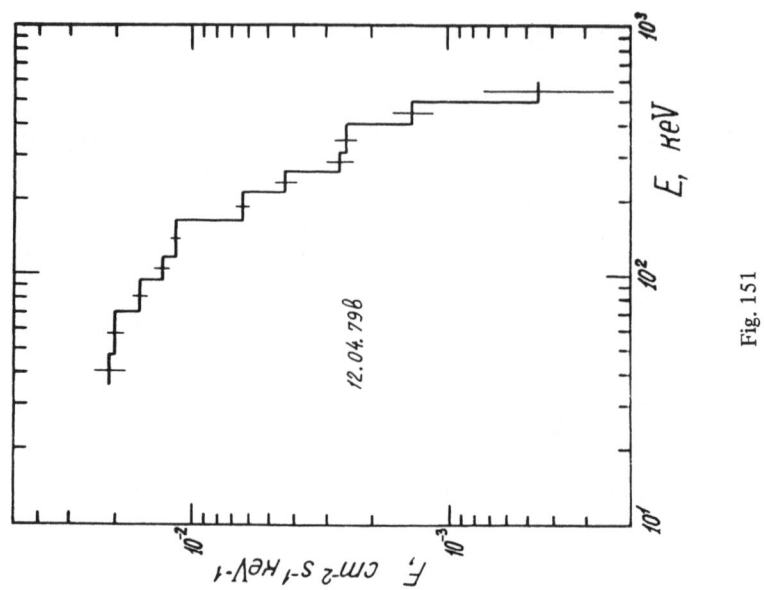

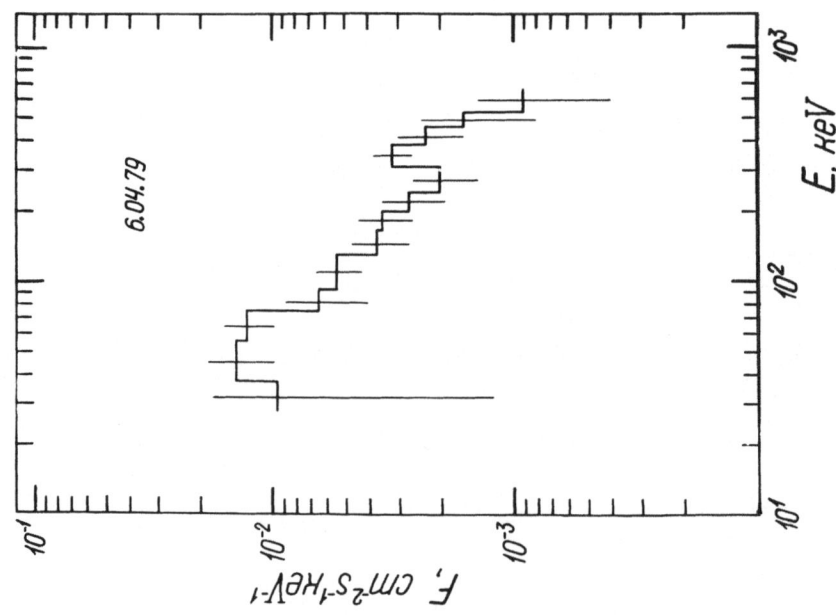

Fig. 151

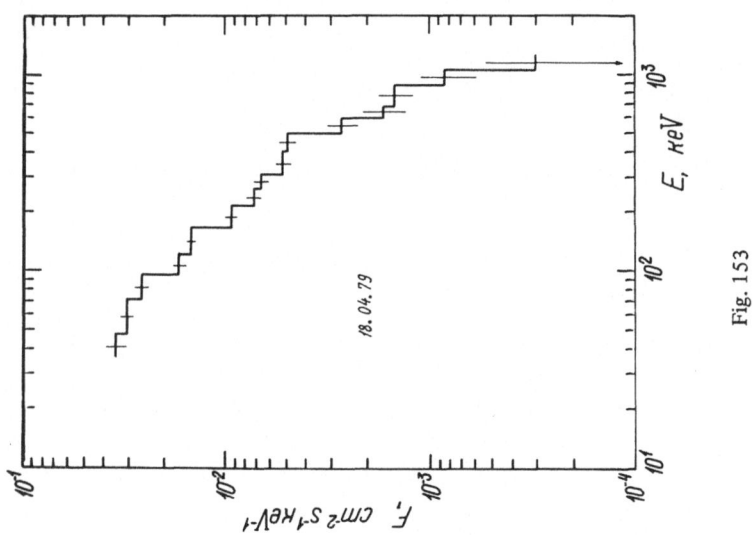

Fig. 153

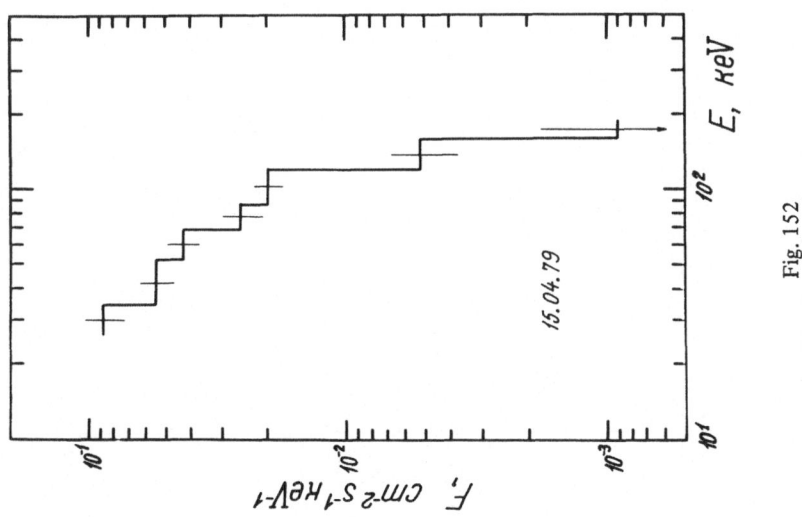

Fig. 152

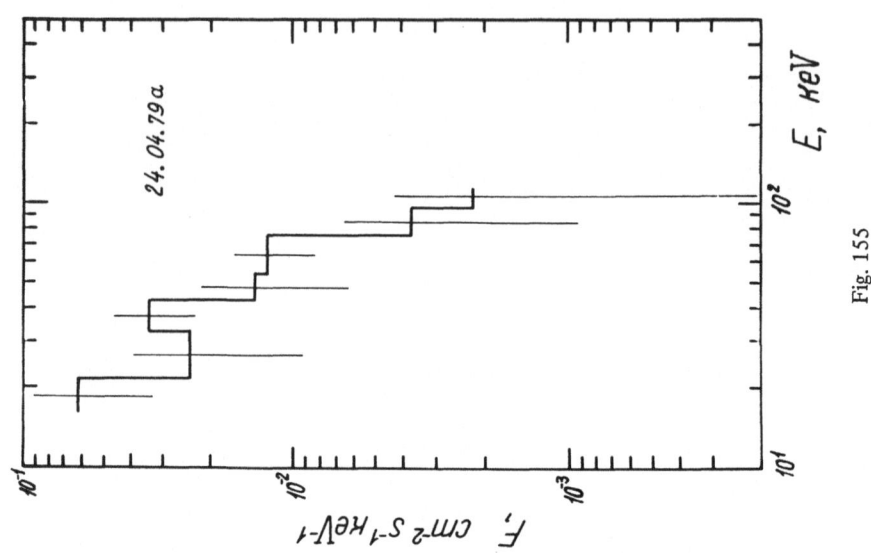

Fig. 155

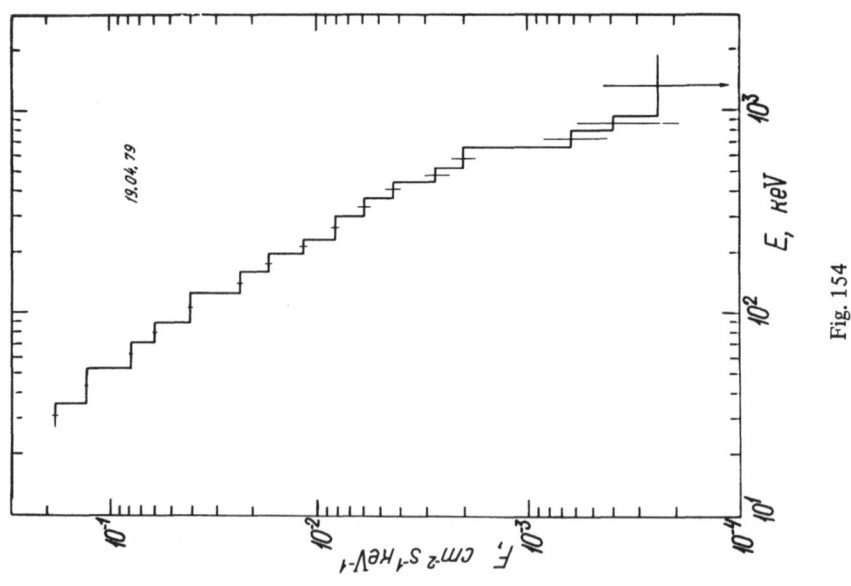

Fig. 154

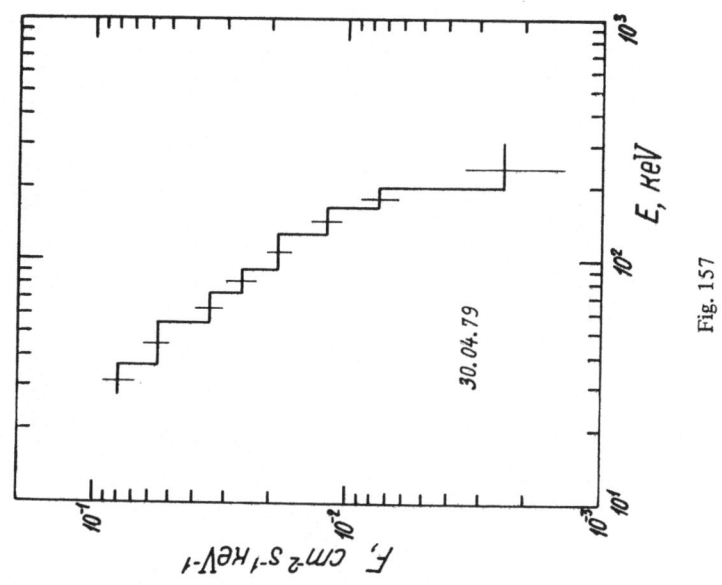

Fig. 157

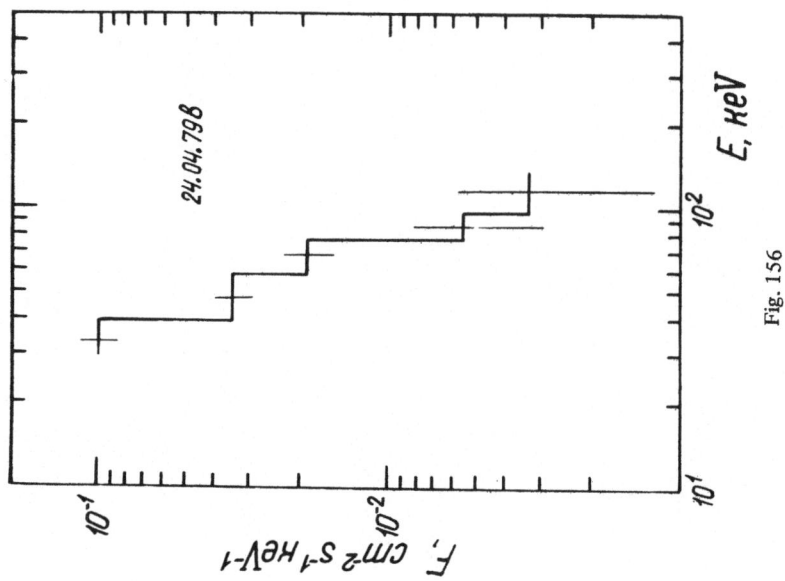

Fig. 156

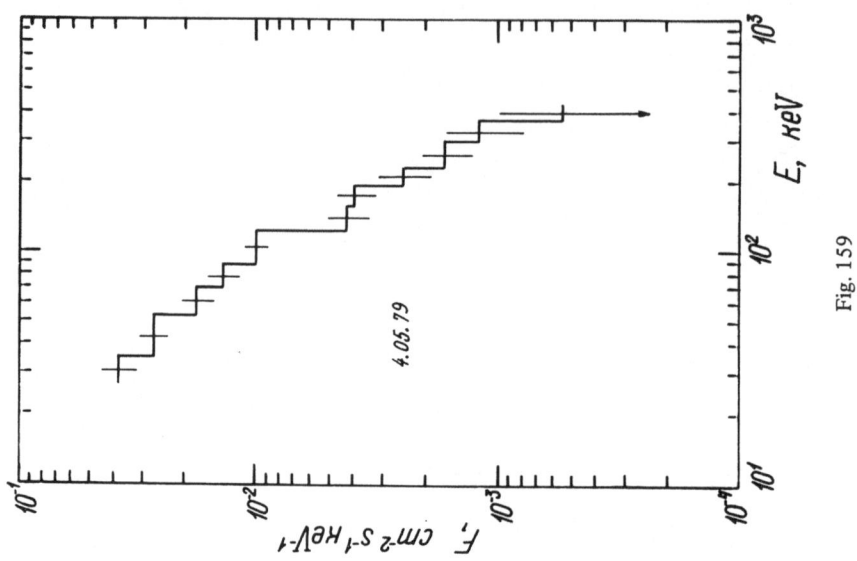

Fig. 159

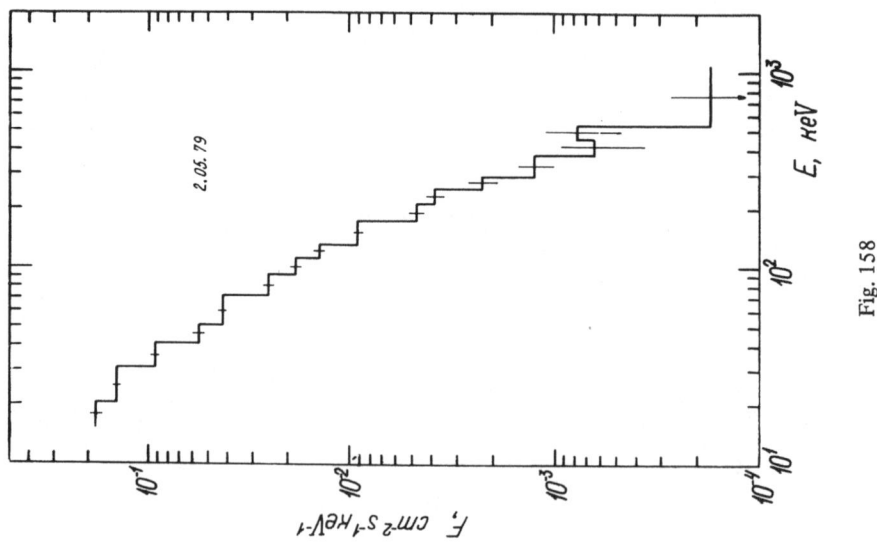

Fig. 158

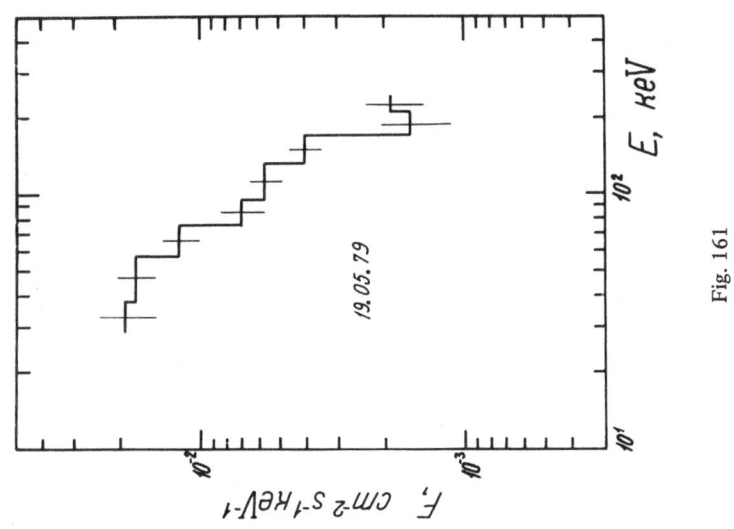

Fig. 161

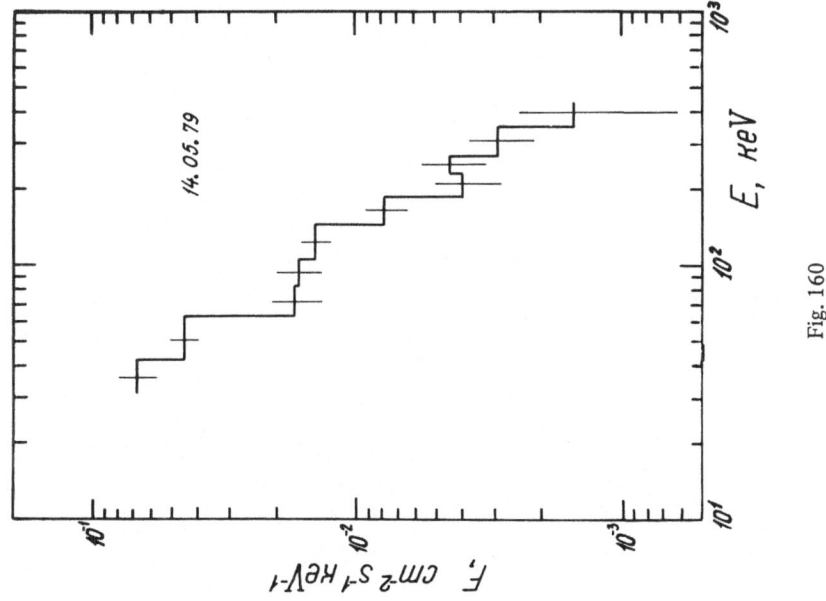

Fig. 160

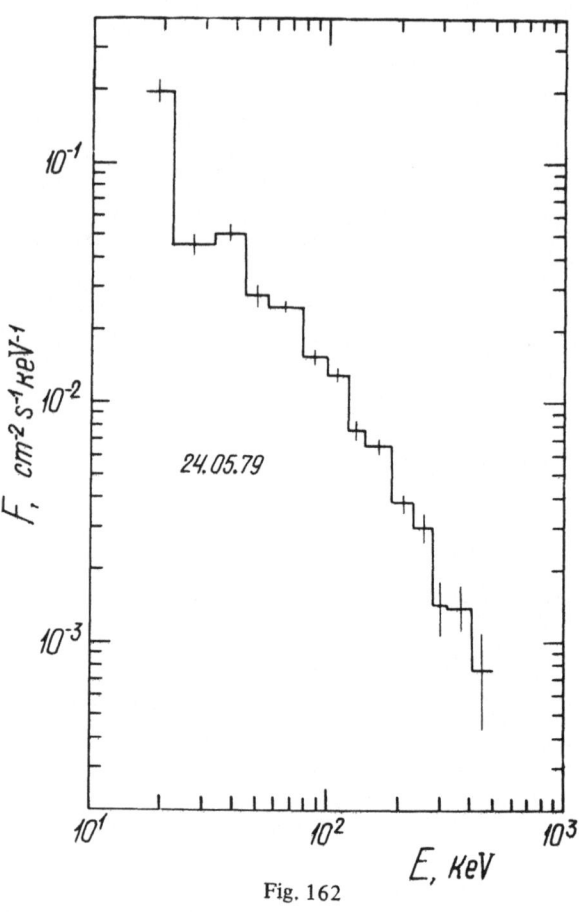

Fig. 162

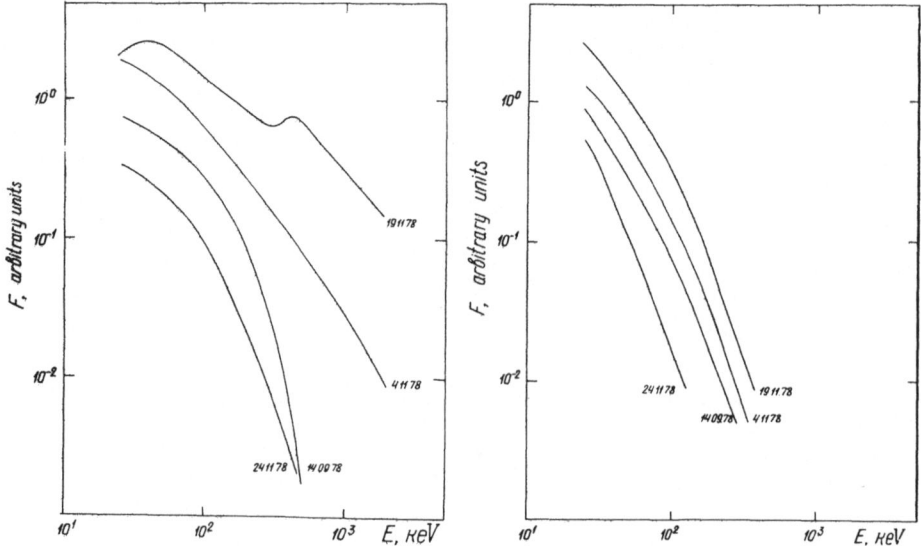

Fig. 163

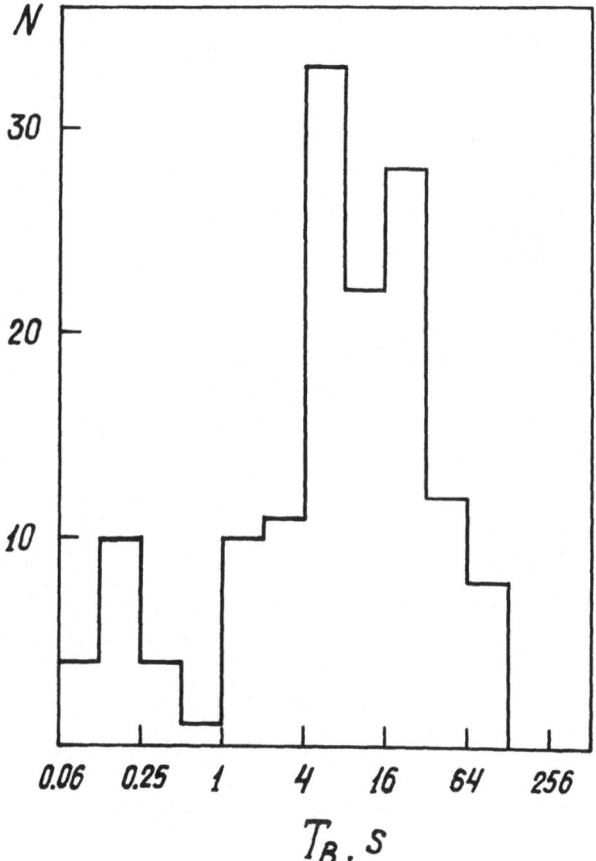

Fig. 164

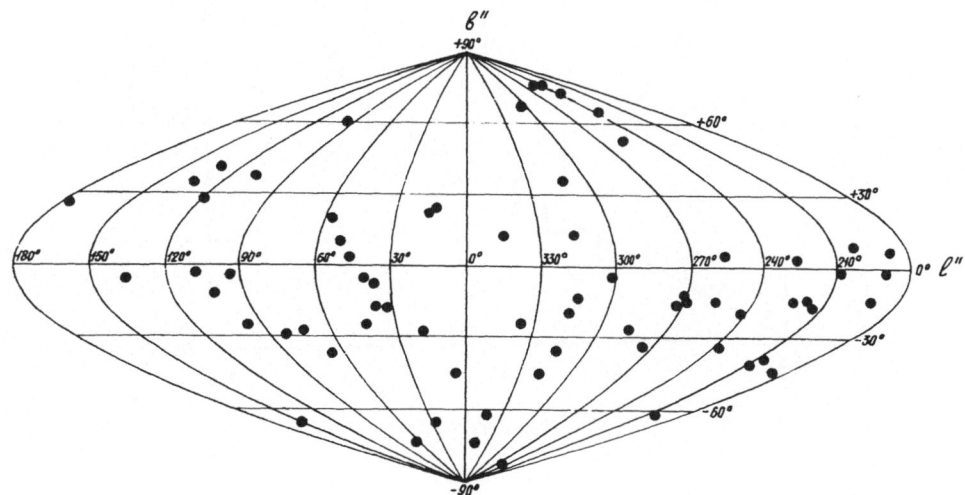

Fig. 165

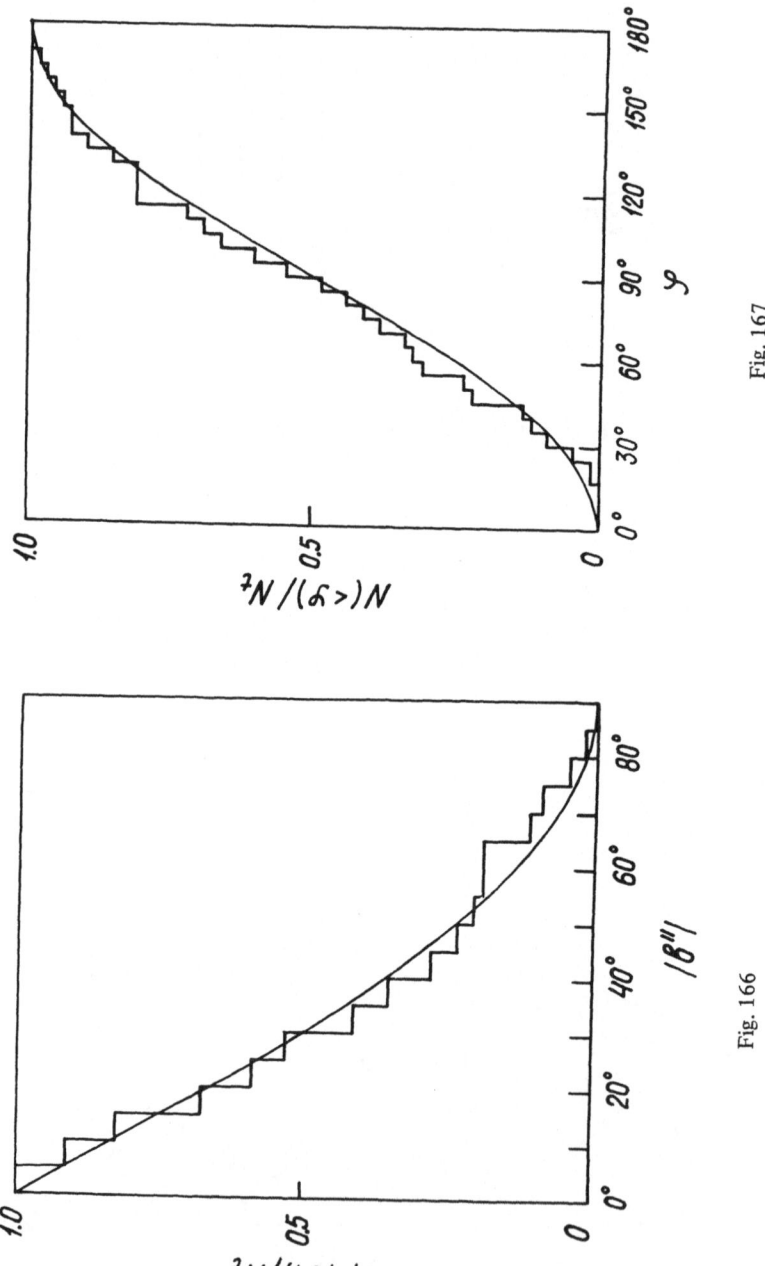

Fig. 167

Fig. 166

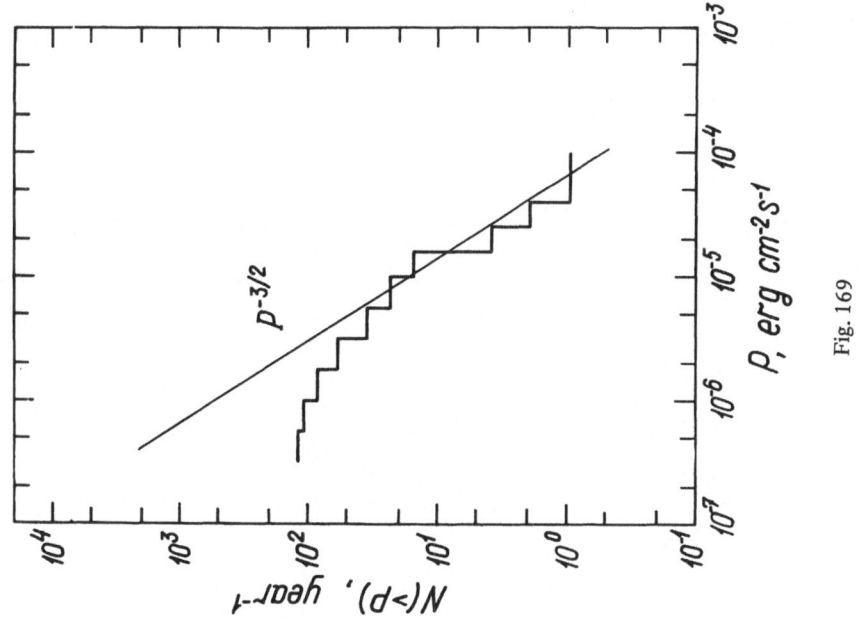

Fig. 169

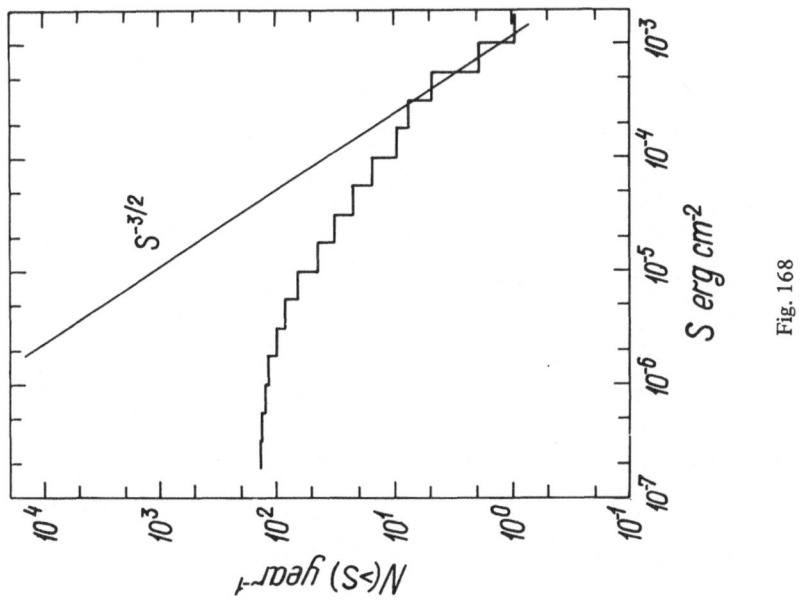

Fig. 168

CATALOG OF COSMIC GAMMA-RAY BURSTS FROM THE KONUS EXPERIMENT DATA
Part III

E. P. MAZETS, S. V. GOLENETSKII, V. N. ILYINSKII, V. N. PANOV,
R. L. APTEKAR, YU. A. GURYAN, M. P. PROSKURA, I. A. SOKOLOV,
Z. YA. SOKOLOVA, and T. V. KHARITONOVA

A. F. Ioffe Physical-Technical Institute, U.S.S.R. Academy of Sciences, Leningrad, U.S.S.R.

and

A. V. DYATCHKOV and N. G. KHAVENSON

Institute of Cosmic Research, U.S.S.R. Academy of Sciences, Moscow, U.S.S.R.

(Received 22 October, 1980)

Abstract. In the third part of the Catalog, the observational data are presented on the gamma-burst studies in the KONUS experiment aboard Venera 11 and Venera 12 spacecraft in the period from the end of May to October 1979. The burst time profiles, energy spectra and source localization data are given.

1. Observations

During the period in question, from the end of May to October 1979 in the KONUS experiment, under the 122-day exposition, 38 gamma bursts were detected. The observational methods and instrumentation have been described in our preceding papers (Mazets and Golenetskii, 1981; Mazets *et al.*, 1979, 1981).

A list of these bursts and their parameters are presented in Table I. Its first column contains the dates of events, which serve further as an event notation. The times T_0 at which the burst detection circuit on Venera 11 and Venera 12 was triggered are given in columns 2 and 3, respectively. Column 4 gives the burst intensity S which is the energy flux in the range $E_\gamma > 30$ keV integrated over the burst duration.

The source direction of a gamma burst is determined in the KONUS experiment both by triangulation from the wave front arrival time difference when an event is recorded on two spacecraft, and independently on each spacecraft from the data of six detectors with an anisotropic angular sensitivity (Mazets and Golenetskii, 1981). Columns 5 and 6 in Table I contain the center position and radius of the triangulation ring on the celestial sphere obtained under the burst detection on two spacecraft. For unambiguously localized sources, columns 7 and 8 give the mean source location in the equatorial and galactic coordinates; the coordinates of the quadrangular localization box are presented in column 9. The equatorial coordinates are given for the epoch 1950.0. In the period in question, unambiguous localization of 21 burst sources has been achieved.

Astrophysics and Space Science **80** (1981) 85–117. 0004–640X/81/0801–0085$04.95.
Copyright © 1981 by D. Reidel Publishing Co., Dordrecht, Holland, and Boston, U.S.A.

TABLE I

Gamma-Burst observations in the KONUS experiment, May to October 1979

Date of event	T₀, h.m.s. UT V-11	V-12	$S(E_\gamma > 30\,\mathrm{keV})$, erg cm⁻²	Ring center α / δ	Ring radius	Mean source position α / δ	l" / b"	Box coordinates α / δ			
(1)	(2)	(3)	(4)	(5)	(6)	(7)	(8)	(9)			
26.05.79	21.50.01	no data	7.4×10^{-6}			331.1 / 6.9	67.0 / -37.5	328.4 / -0.6	339.6 / 3.8	322.0 / 9.9	333.7 / 14.4
29.05.79	20.23.01	"	3.1×10^{-6}								
4.06.79	06.34.04	"	1.6×10^{-5}			326.1 / -76.3	314.5 / -36.4	298.3 / -75.8	337.0 / -83.5	322.2 / -68.6	349.9 / -73.7
5.06.79	17.45.02	"	5.4×10^{-6}			356.9 / 50.1	112.4 / -11.3	345.6 / 53.2	2.6 / 55.5	352.7 / 44.2	7.1 / 45.6
10.06.79	05.50.04	"	6.1×10^{-6}			111.4 / -11.3	226.5 / 2.8	116.1 / -14.7	106.9 / -13.5	115.8 / -9.1	106.9 / -8.1
12.06.79	04.42.54	"	3.0×10^{-6}			51.6 / -1.9	185.0 / -44.5	49.7 / -6.4	57.7 / -3.9	45.8 / 1.1	53.6 / 2.6
13.06.79	14.06.54	"	4.0×10^{-7}			252.4 / 87.2	119.5 / 28.8	245.4 / 74.6	151.5 / 78.0	309.5 / 76.2	62.9 / 79.9
19.06.79	11.36.58	"	2.0×10^{-7}								
22.06.79	00.41.30	"	7.2×10^{-5}			321.3 / -37.7	5.3 / -46.4	319.6 / -40.6	325.7 / -39.1	317.2 / -36.1	322.9 / -34.7
28.06.79a	20.19.37	"	8.7×10^{-6}								
28.06.79b	22.54.07	"	5.0×10^{-7}								
5.07.79	0.40 ÷ 3.15	"	1.5×10^{-6}								
6.07.79	12.21.19	"	6.0×10^{-7}								
9.07.79	17.00.50	"	4.7×10^{-7}								
12.07.79	04.05.09	"	1.0×10^{-6}			343.6 / -27.3	26.0 / -64.7	341.1 / -34.6	353.4 / -29.9	334.8 / -24.0	346.1 / -19.9
20.07.79	21.21.11	21.22.29	4.6×10^{-6}	92.732 / 25.742	61.286 ± 0.040	76.27 / -33.51	235.6 / -35.3	80.39 / -34.46	80.50 / -34.39	72.21 / -32.38	72.32 / -32.33
28.07.79	02.44.00	02.44.35	1.4×10^{-6}	98.000 / 25.762	78.475 ± 0.015	23.531 / -5.320	150.4 / -65.4	26.780 / -11.254	26.809 / -11.242	20.468 / 0.676	20.496 / 0.687

Date											
31.07.79	10.49.55	10.53.03	8.1×10^{-5}	100.514 / 25.689	$0.66 \pm {}^{2.7}_{0.66}$	98.10 / 25.83	187.4 / 8.2	95.0 / 23.3	100.8 / 23.0	95.3 / 28.6	101.5 / 28.3
10.08.79	no data	04.12.14	7.0×10^{-7}								
18.08.79	,,	11.33.56	3.0×10^{-7}								
26.08.79	18.31.04	18.32.31	1.0×10^{-5}	122.055 / 22.990	68.588 ± 0.026	109.602 / -44.646	255.7 / -14.0	102.728 / -43.280	102.724 / -43.224	116.729 / -45.452	116.710 / -45.399
29.08.79	08.05.40	08.06.28	3.2×10^{-6}	124.316 / 22.498	78.514 ± 0.060	122.93 / -56.01	270.0 / -11.9	110.06 / -55.12	110.03 / -55.00	135.94 / -55.45	135.82 / -55.34
10.09.79	no data	03.24.45	3.9×10^{-6}			62.4 / -9.3	201.3 / -39.5	58.0 / -14.2	66.0 / -12.4	58.7 / -6.2	66.3 / -4.6
13.09.79	03.35.00	no data	1.0×10^{-5}								
24.09.79	no data	10.36.06	5.0×10^{-6}								
25.09.79	09.34.45	not detected	5.6×10^{-6}			117.1 / -59.9	272.0 / -16.5	111.1 / -61.4	119.4 / -62.9	115.2 / -56.9	122.6 / -58.2
30.09.79	16.46.45	no data	1.7×10^{-5}								
3.10.79	11.40.19	,,	8.0×10^{-6}			166.4 / -1.6	257.6 / 51.9	160.6 / -6.3	167.3 / -9.6	165.9 / 6.2	172.5 / 3.0
6.10.79	08.10.26	,,	7.3×10^{-6}			89.8 / 19.4	189.5 / -1.5	87.7 / 14.2	93.3 / 14.2	86.5 / 24.5	92.4 / 24.6
7.10.79	16.15.27	,,	3.0×10^{-6}								
14.10.79	11.11.26	11.12.26	1.4×10^{-5}	169.858 / 5.419	79.747 ± 0.087	96.03 / -33.78	241.0 / -19.6	96.60 / -36.01	96.83 / -36.02	95.30 / -31.56	95.52 / -31.56
16.10.79	17.01.26	17.02.50	8.0×10^{-7}	172.327 / 4.213	75.543 ± 0.087	143.82 / 78.30	132.9 / 34.8	114.90 / 70.54	115.43 / 70.49	202.78 / 78.17	202.86 / 77.95
21.10.79	no data	11.46.12	4.0×10^{-7}								
23.10.79	11.00.15	10.59.26	9.0×10^{-7}	359.922 / -4.454	82.010 ± 0.017	80.775 / 11.939	191.4 / -12.8	79.484 / 21.863	79.448 / 21.860	81.764 / 1.987	81.730 / 1.983
30.10.79a	no data	11.31.52	8.0×10^{-6}								
30.10.79b	,,	17.00.46	3.3×10^{-6}								
31.10.79a	09.18.33	09.19.18	6.5×10^{-6}	189.188 / -4.151	82.720 ± 0.081	254.8 / -82.3	310.7 / -23.6	233.5 / -85.7	263.3 / -78.5	231.8 / -85.6	262.5 / -78.5
31.10.79b	no data	14.20.34	1.5×10^{-6}								

In the whole, during the period from September 1978 to October 1979 inclusive covering 339 days of observations the KONUS instruments detected 123 bursts.

2. Temporal Structure of Bursts

The time profiles of the gamma bursts detected are shown in Figures 1–40. The graphs show the count rate in the energy window 50–150 keV vs. time with 0.25 s resolution (after $T - T_0 \approx 34$ s – with 1 s resolution). Separate points before the unbroken histogram represent the prehistory of the gamma burst recorded with the same detector. The horizontal dashed line shows the mean background level measured before and after a gamma burst. In cases of sufficient statistical reliability, the initial burst stage is shown with 1/64 s resolution. For short bursts, only 1/64 s resolution time profiles are presented.

3. Energy Spectra

The differential energy spectra of gamma bursts are presented in Figures 41–73. Eight energy spectra are recorded for each burst during eight consecutive time intervals of 4 s each. Only spectra summarized over these intervals are usually presented. In the cases when a substantial evolution of spectrum in time is revealed, we present some of the spectra obtained in several 4 s time intervals after T_0. The numbers of these intervals are labelled near the corresponding curves in the graphs. For several weaker bursts the spectral data have a poor statistics; such data are not presented here.

References

Mazets, E. P. and Golenetskii, S. V.: 1981, *Astrophys. Space Sci.* **75**, 47.

Mazets, E. P., Golenetskii, S. V., Ilyinskii, V. N., Panov, V. N., Aptekar, R. L., Guryan, Yu. A., Sokolov, I. A., Sokolova, Z. Ya, and Kharitonova, T. V.: 1979, *Kosmicheskiye Issledoviniya* **17**, 812.

Mazets, E. P., Golenetskii, S. V., Ilyinskii, V. N., Panov, V. N., Aptekar, R. L., Guryan, Yu. A., Proskura, M. P., Sokolov, I. A., Sokolova, Z. Ya., Kharitonova, T. V., Dyatchkov, A. V. and Khavenson, N. G.: 1981, *Astrophys. Space Sci.* **80**, 3 (this issue).

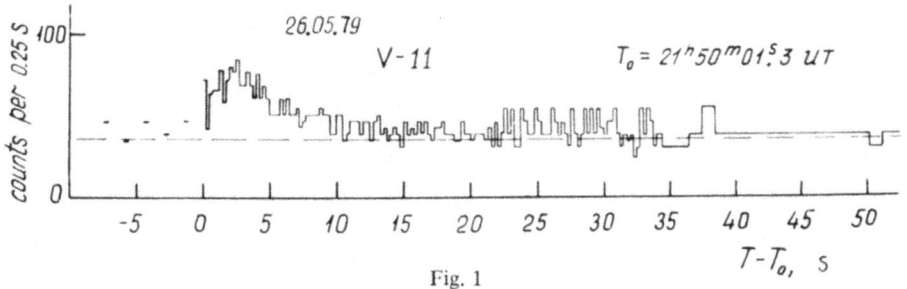

Fig. 1

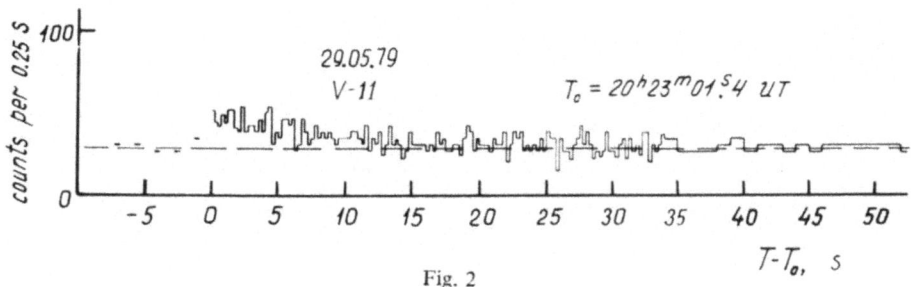

Fig. 2

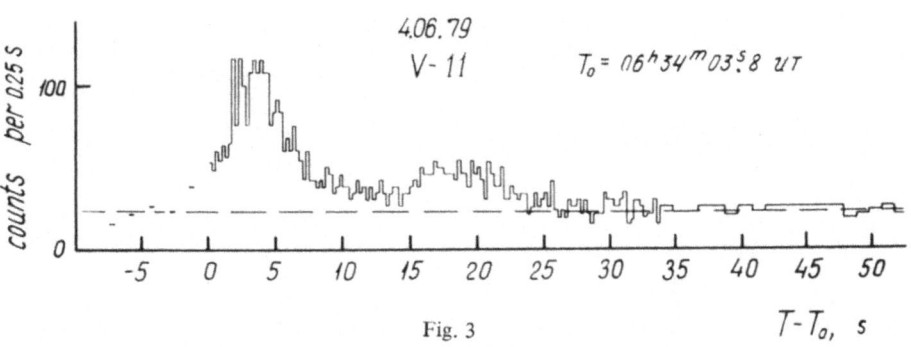

Fig. 3

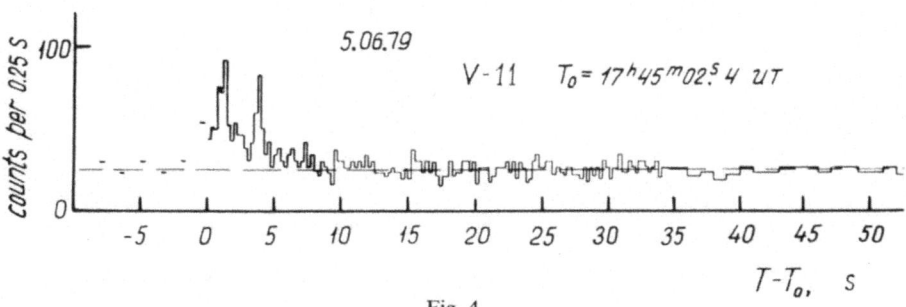

Fig. 4

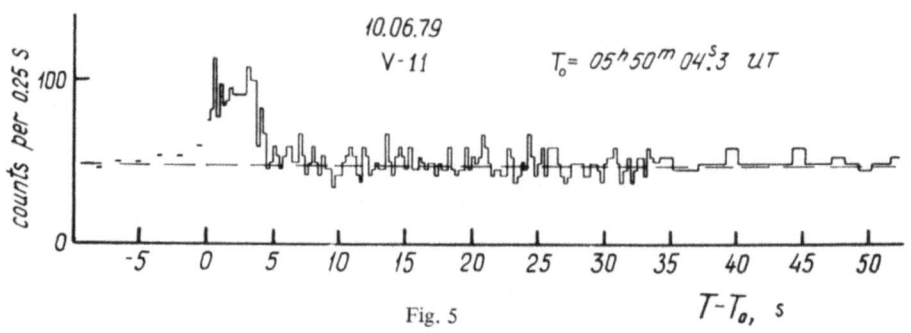

Fig. 5

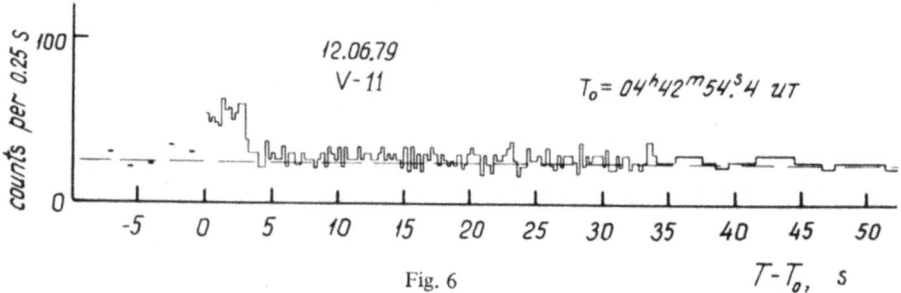

Fig. 6

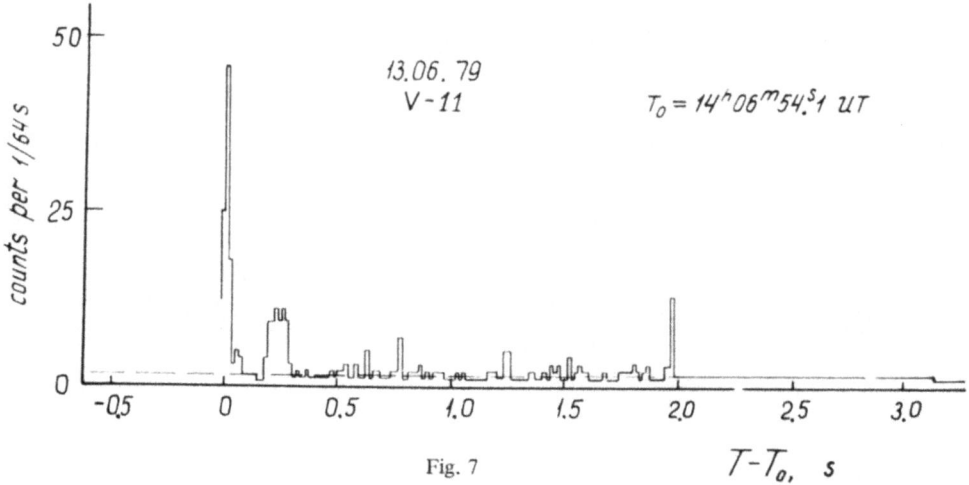

Fig. 7

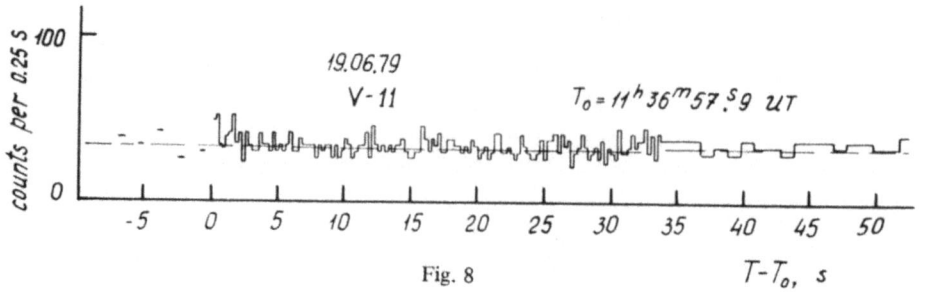

Fig. 8

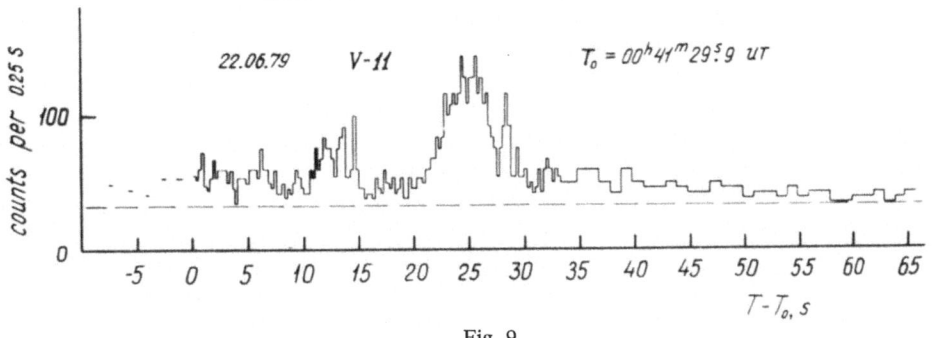

Fig. 9

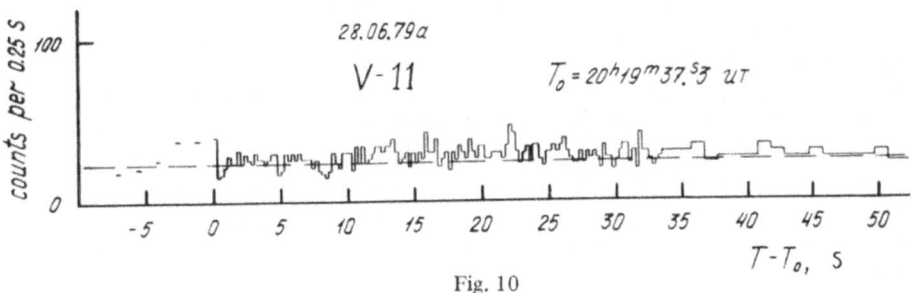

Fig. 10

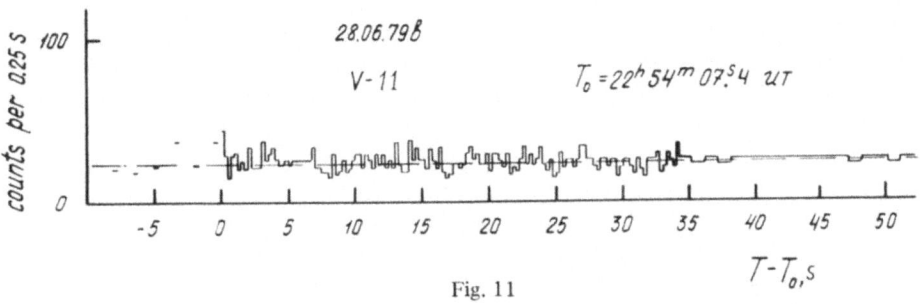

Fig. 11

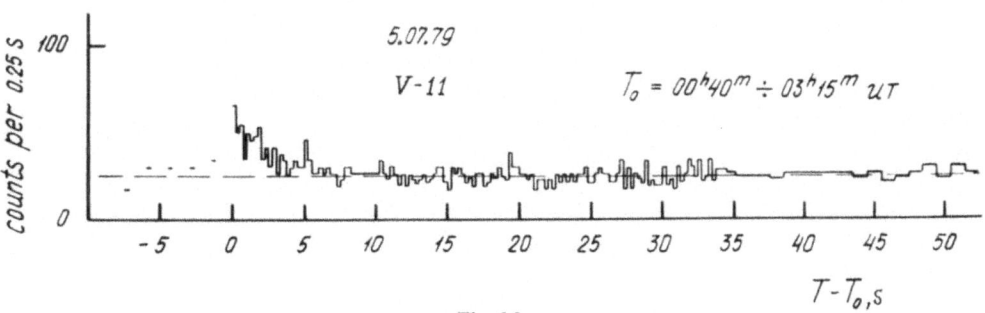

Fig. 12

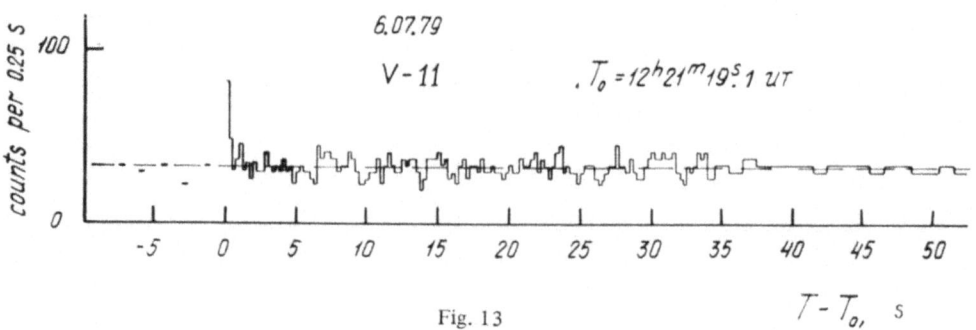

Fig. 13

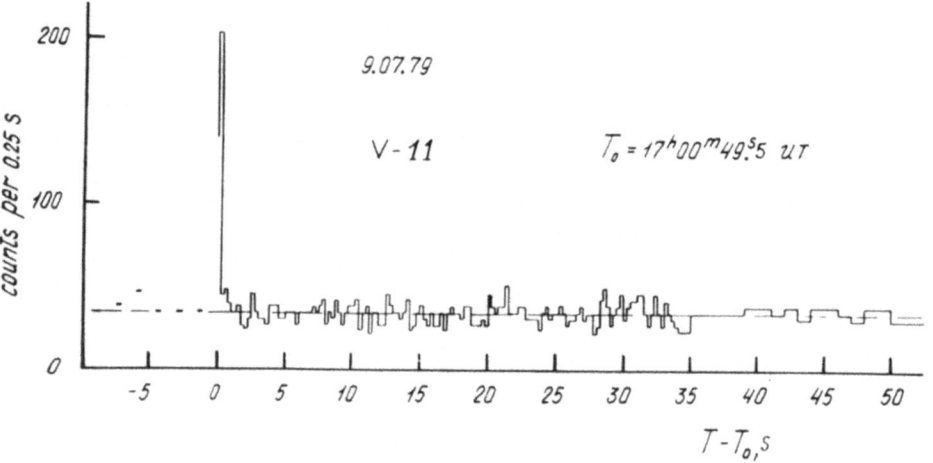

Fig. 14

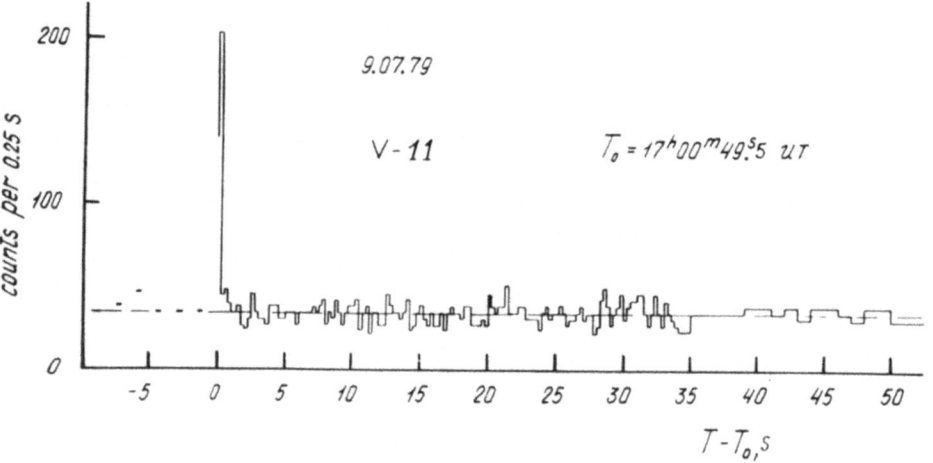

Fig. 15

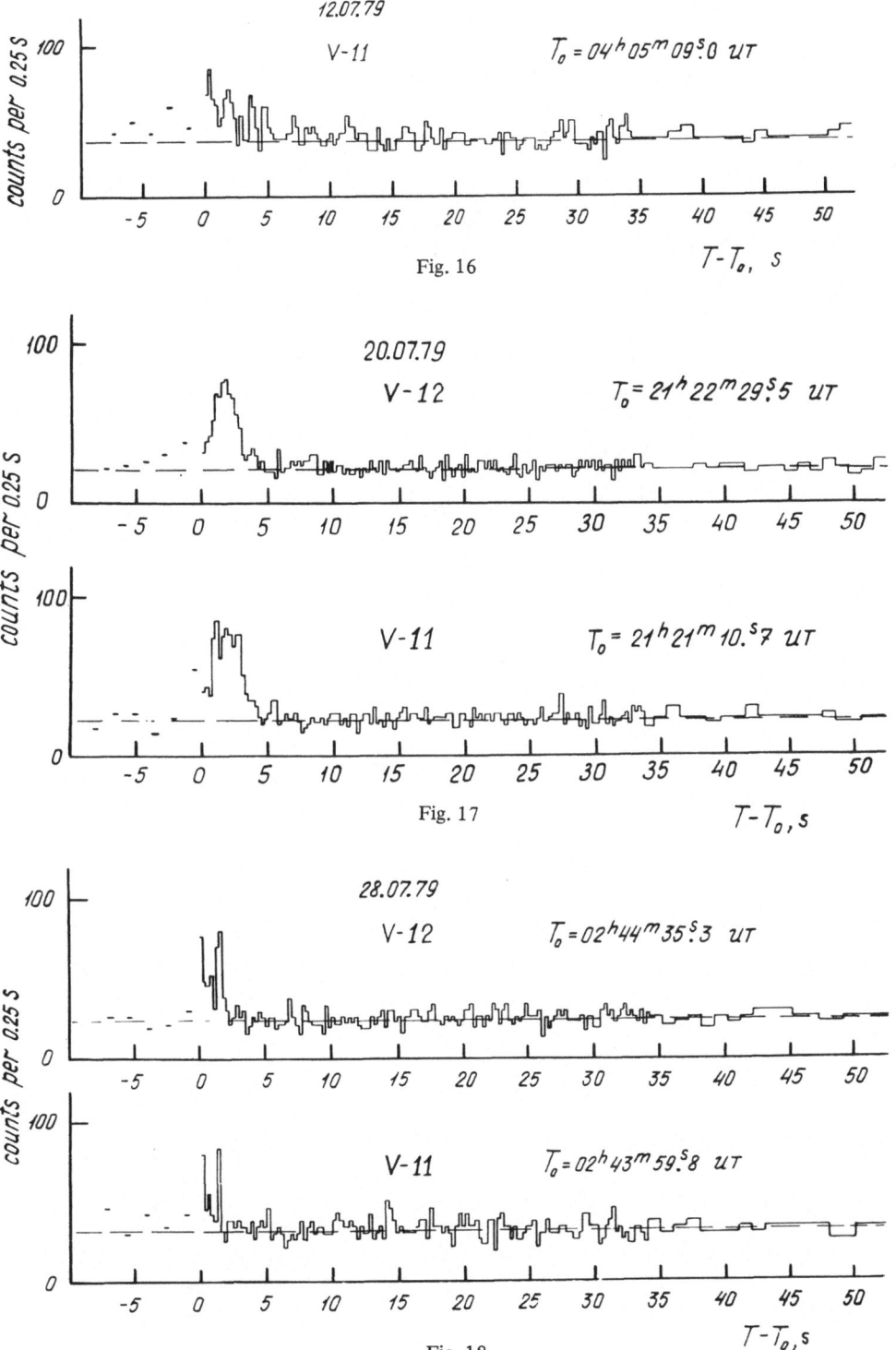

Fig. 16

Fig. 17

Fig. 18

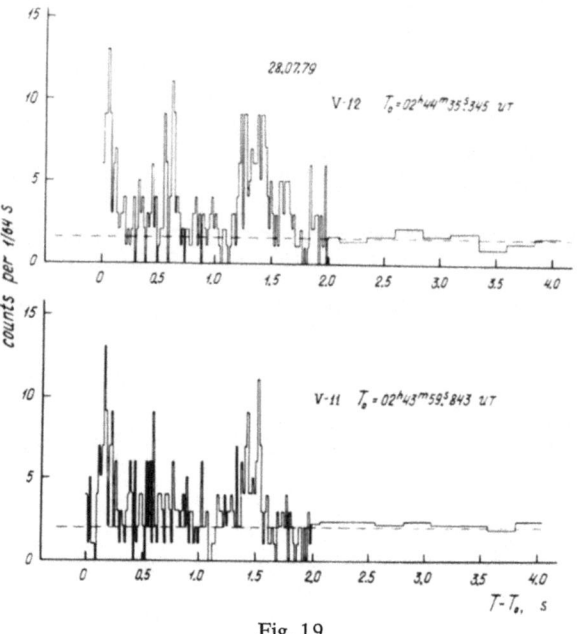

Fig. 19

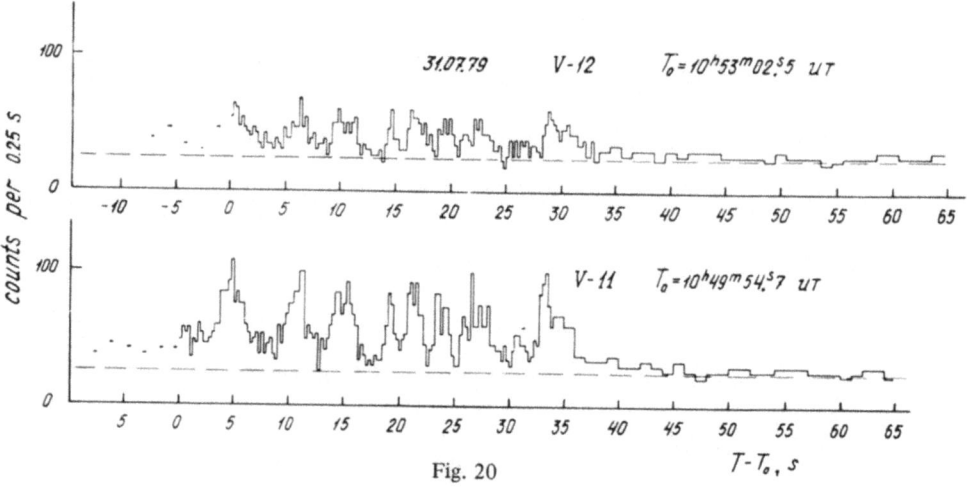

Fig. 20

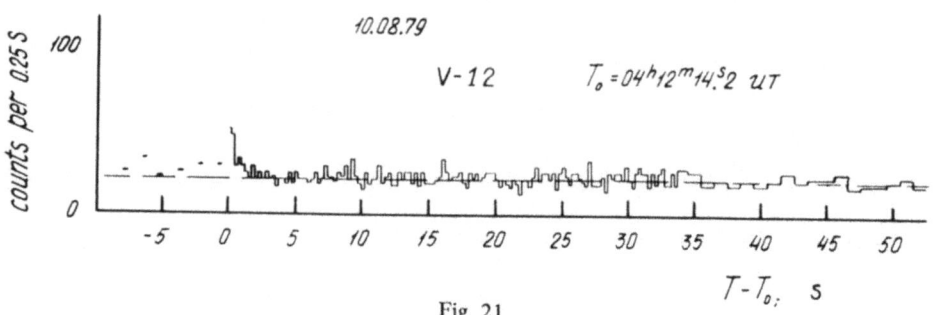

Fig. 21

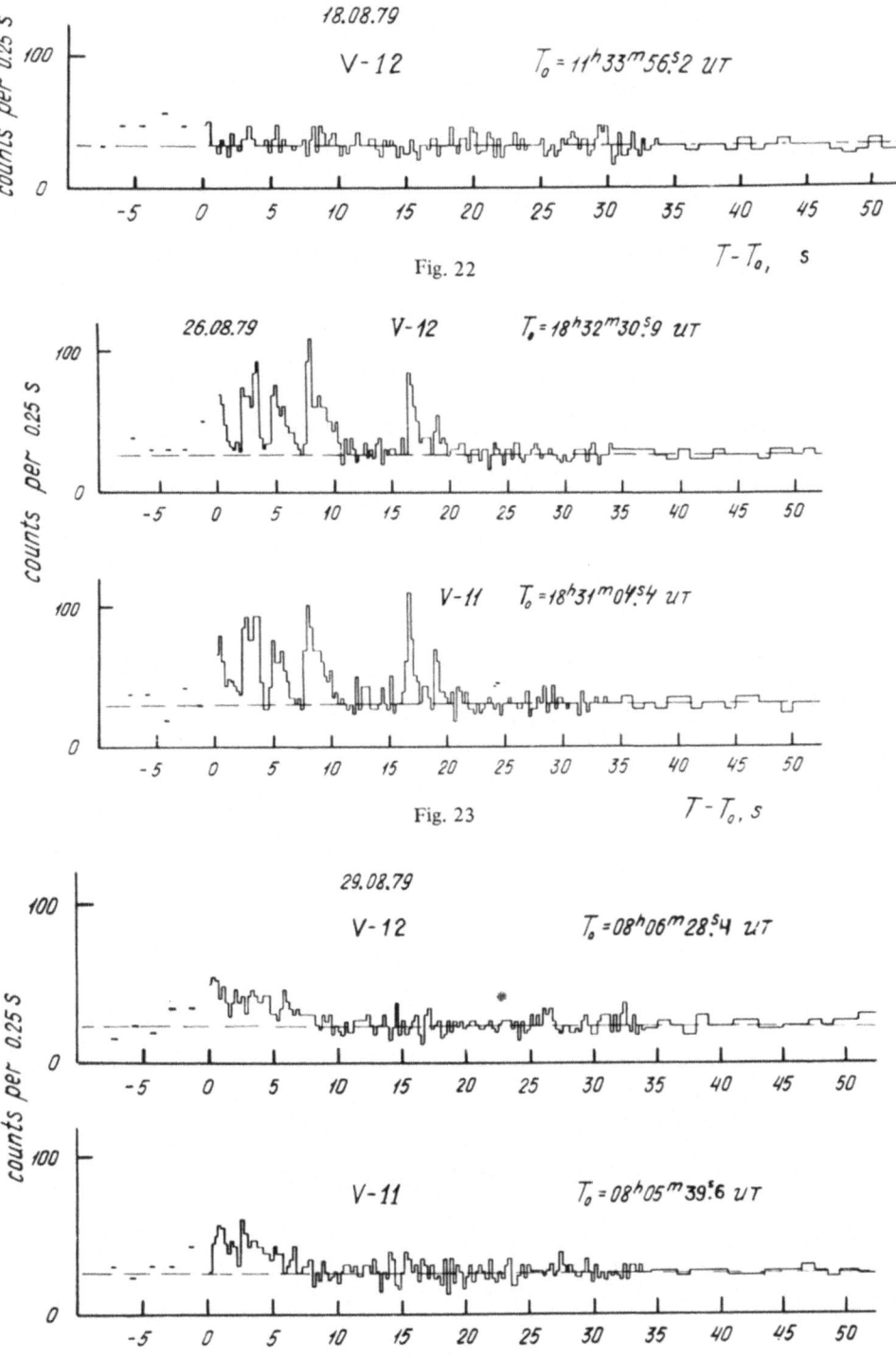

18.08.79

V-12 $T_o = 11^h33^m56^s2$ UT

Fig. 22 $T - T_o$, s

26.08.79   V-12 $T_o = 18^h32^m30^s9$ UT

V-11 $T_o = 18^h31^m04^s4$ UT

Fig. 23 $T - T_o$, s

29.08.79

V-12 $T_o = 08^h06^m28^s4$ UT

V-11 $T_o = 08^h05^m39^s6$ UT

Fig. 24 $T - T_o$, s

counts per 0.25 s

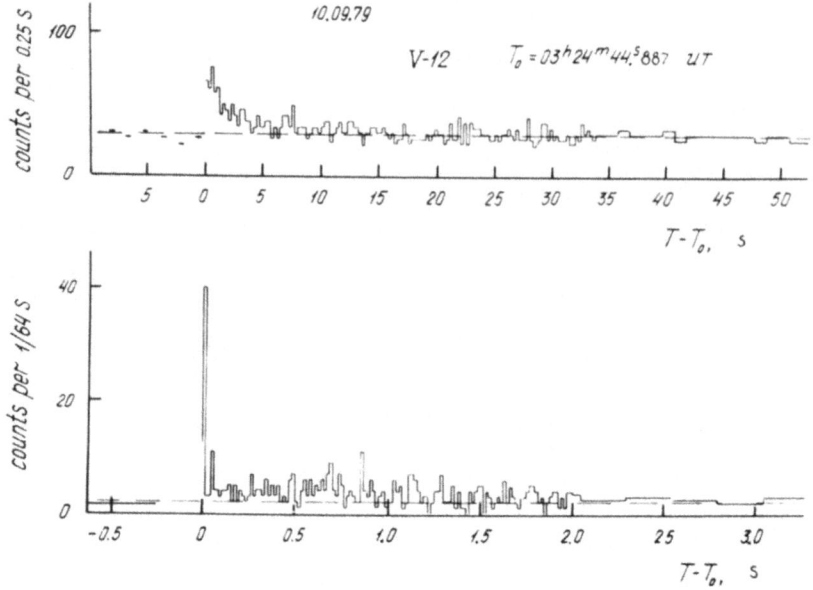

Fig. 25

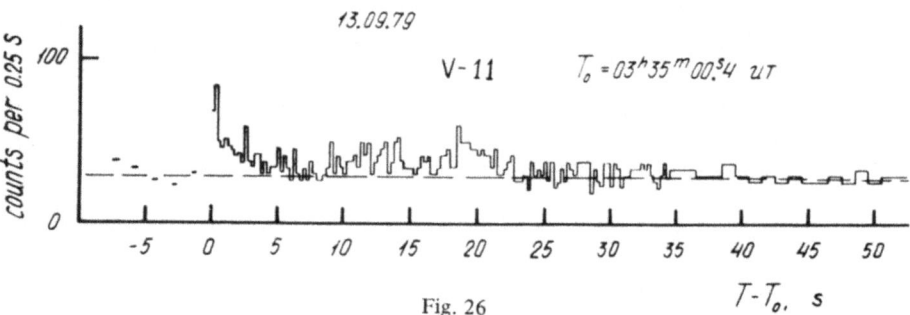

Fig. 26

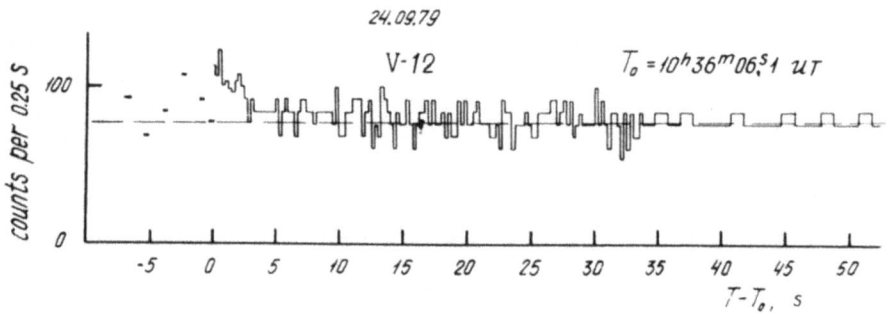

Fig. 27

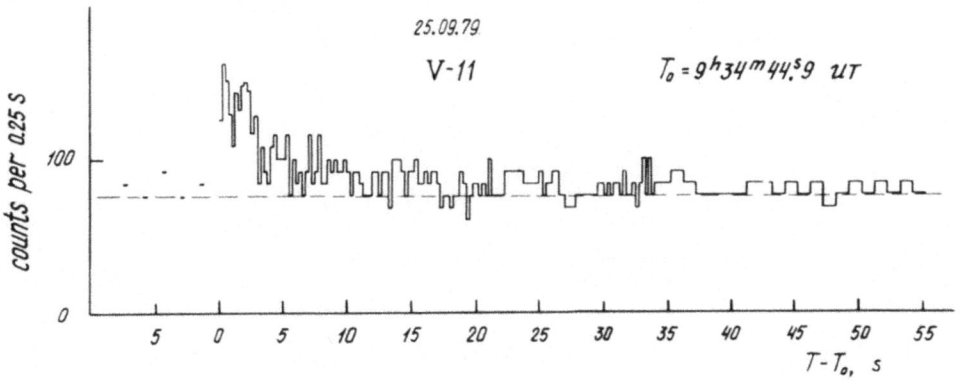

Fig. 28

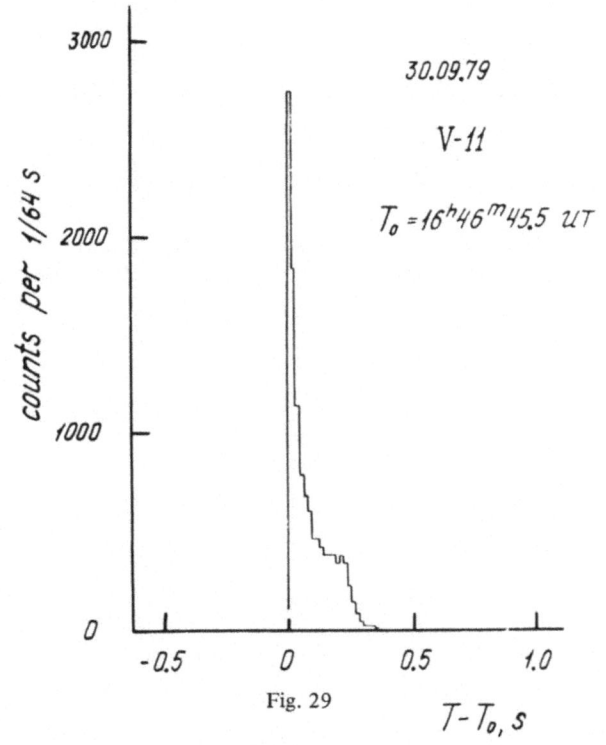

Fig. 29

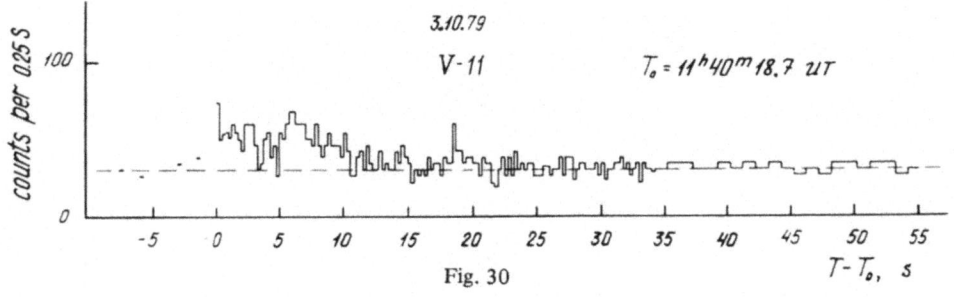

Fig. 30

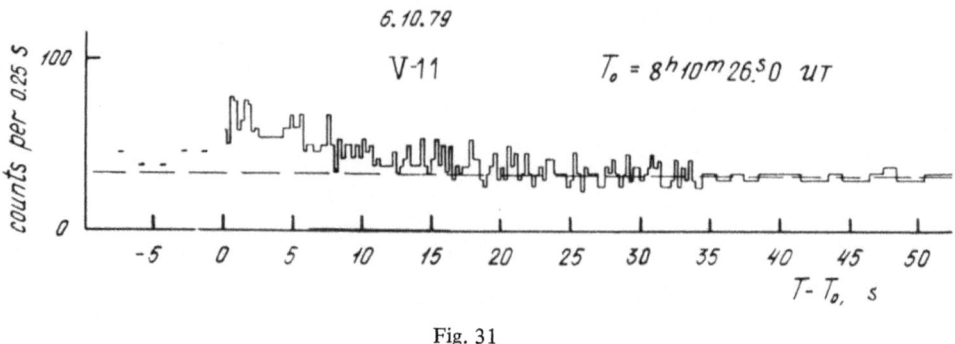

Fig. 31

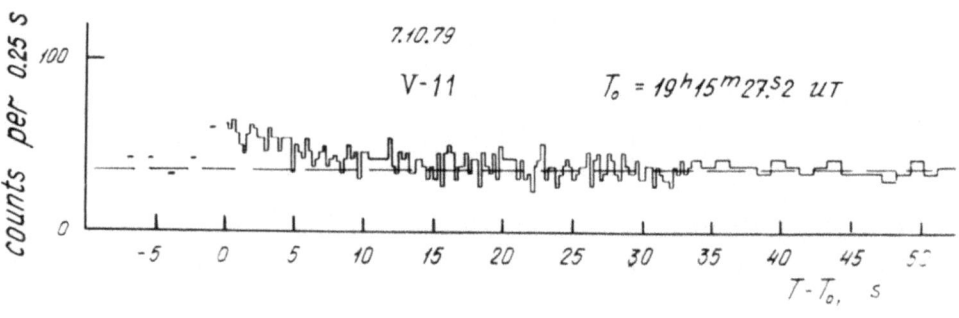

Fig. 32

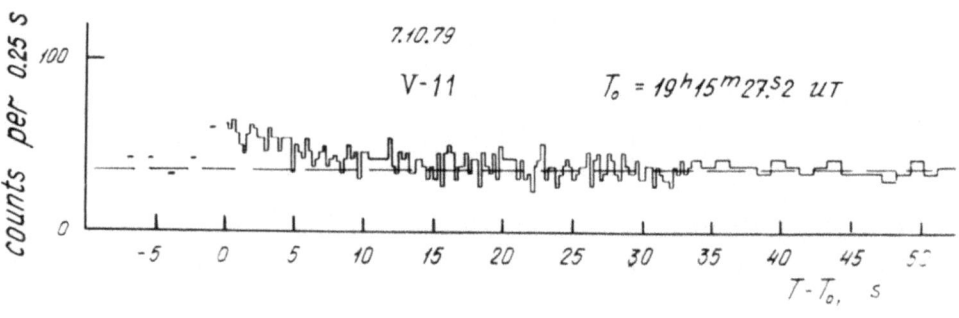

Fig. 33

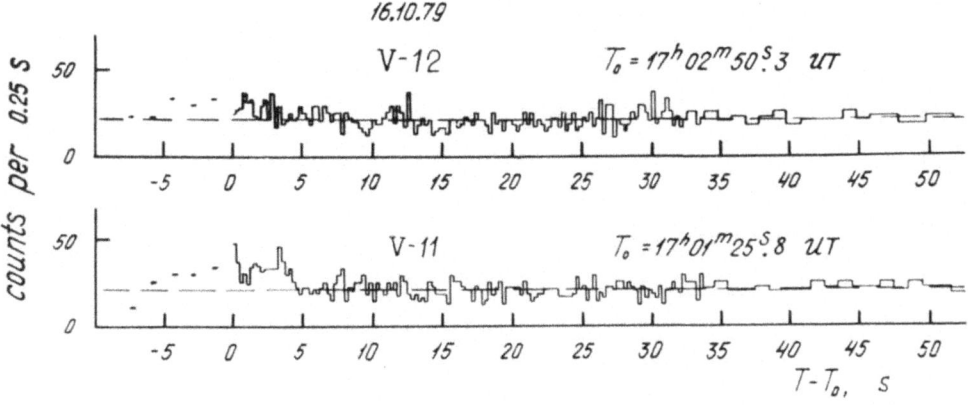

Fig. 34

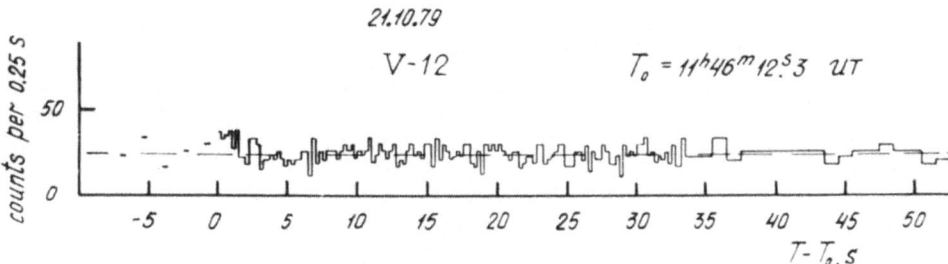

Fig. 35

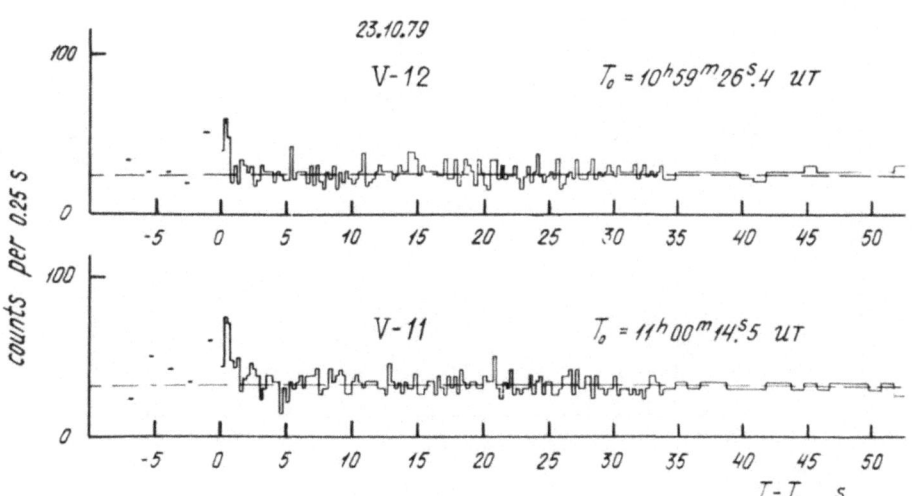

Fig. 36

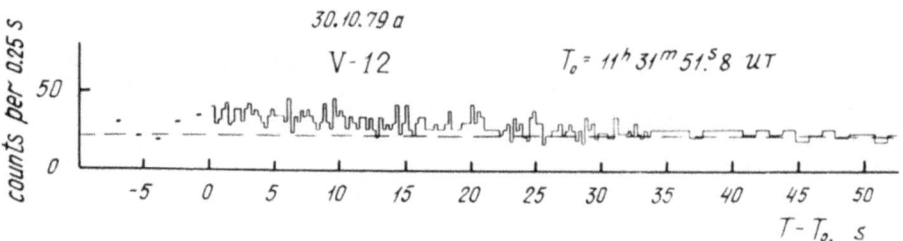

Fig. 37

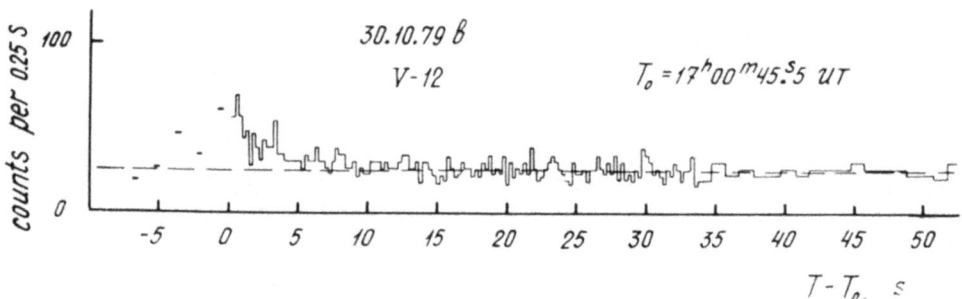

Fig. 38

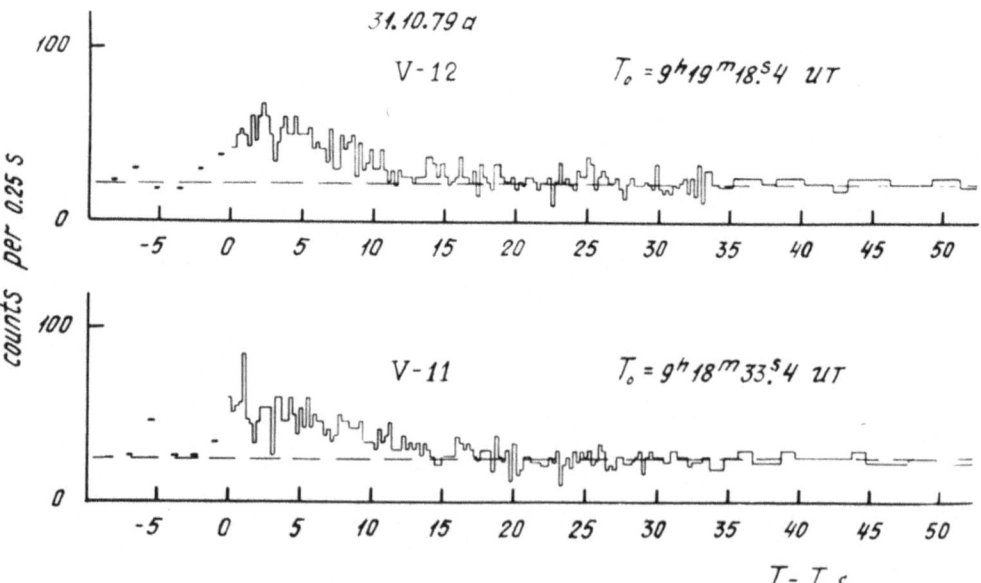

Fig. 39

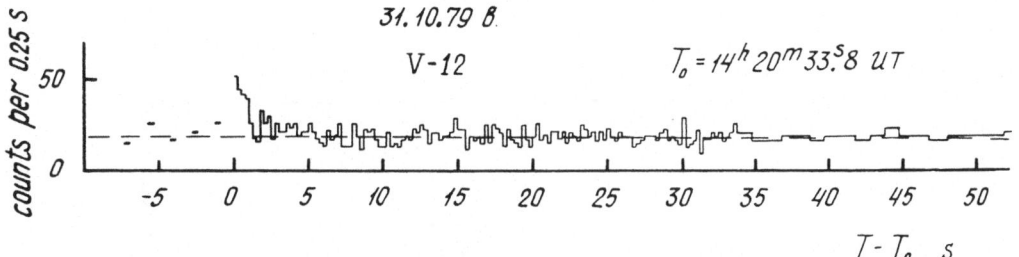

31.10.79 B.

V-12 $T_o = 14^h 20^m 33\overset{s}{.}8$ UT

Fig. 40

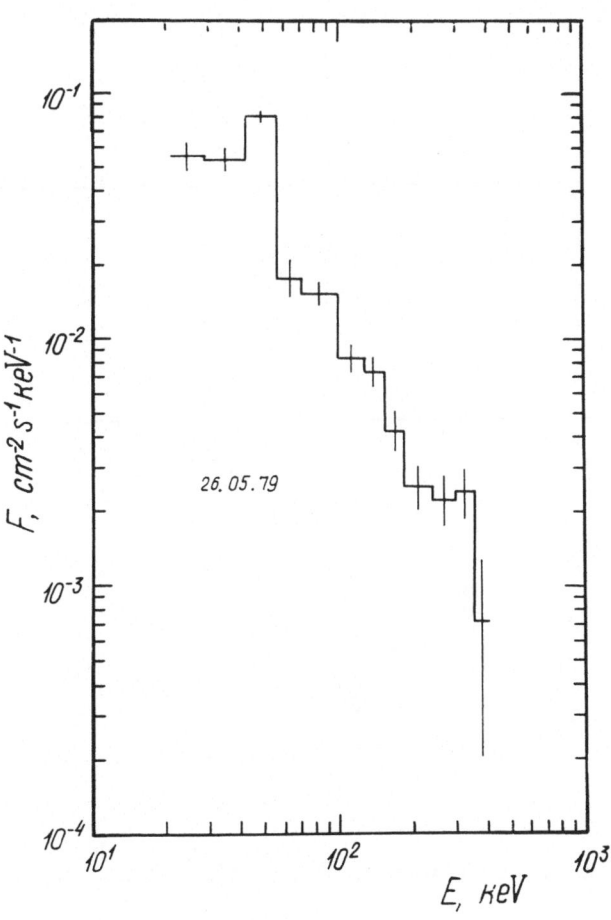

26.05.79

Fig. 41

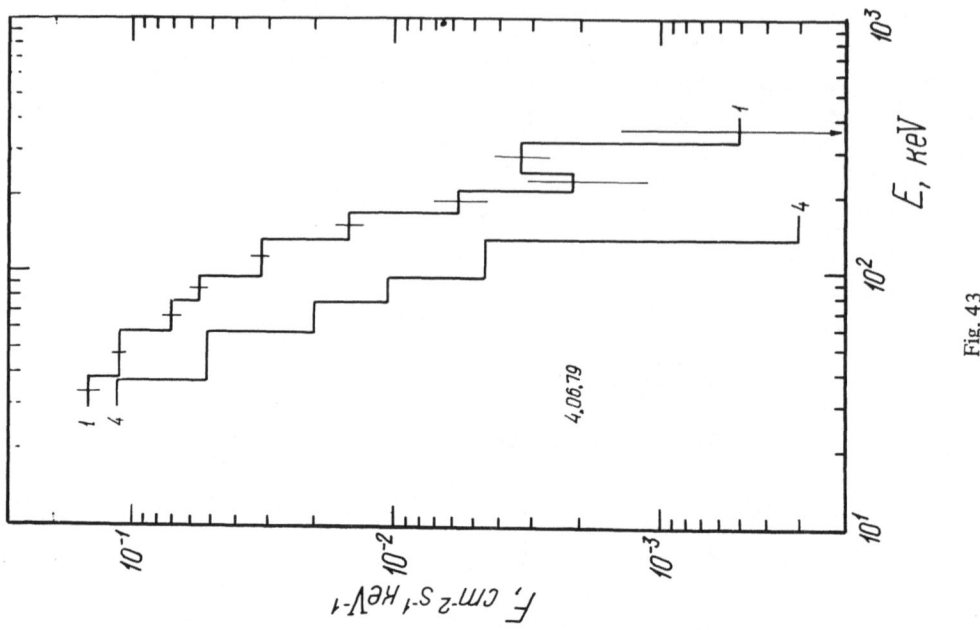

Fig. 43

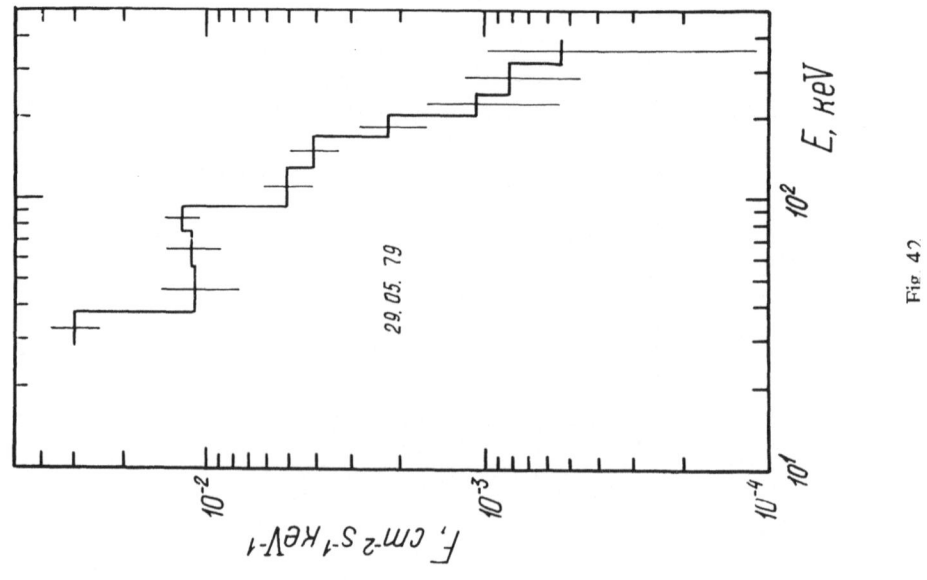

Fig. 42

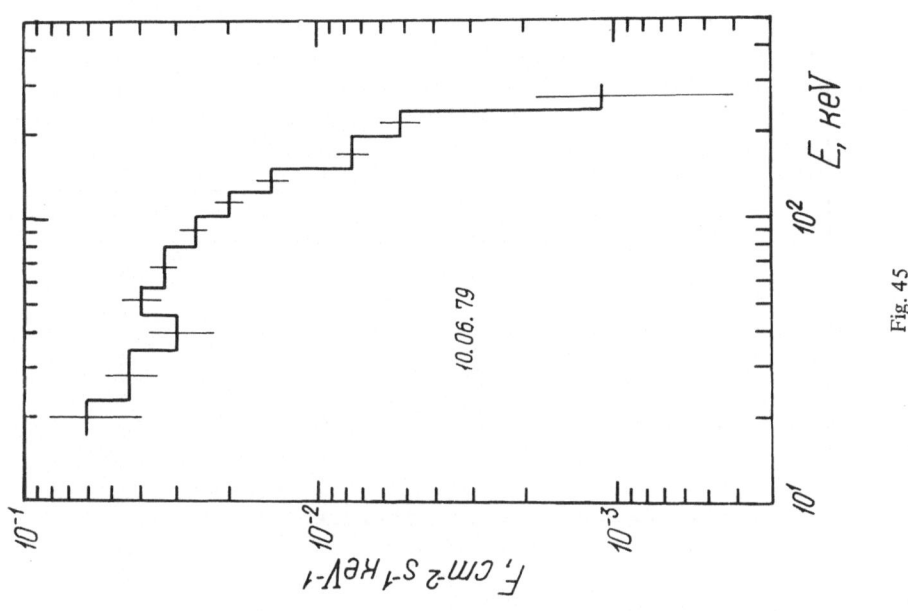

Fig. 45

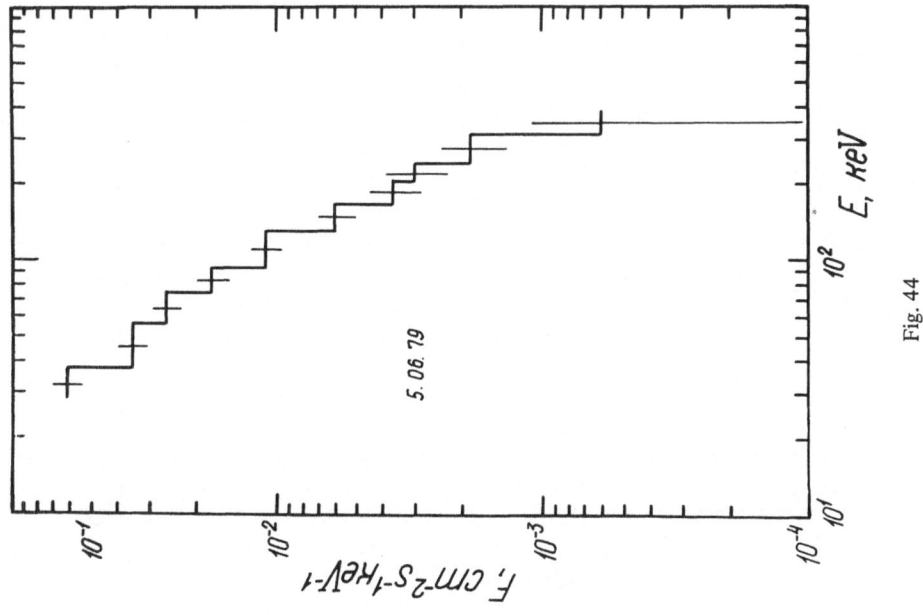

Fig. 44

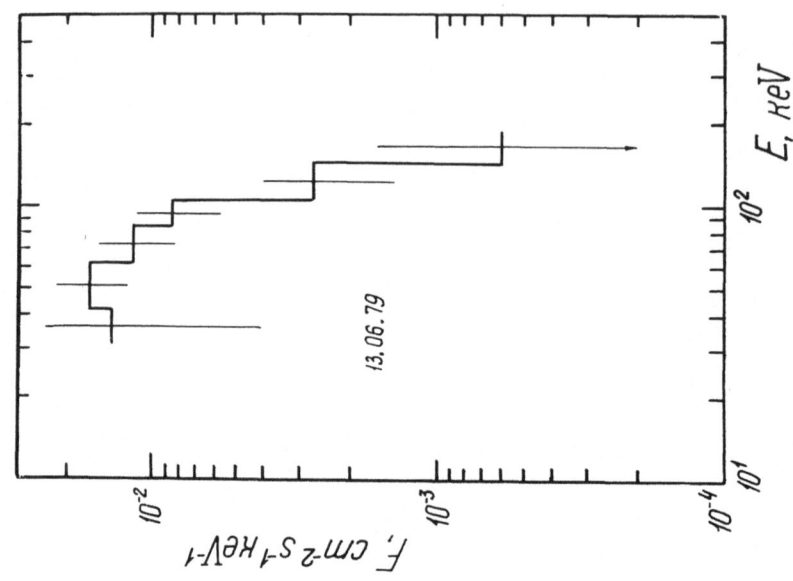

Fig. 47

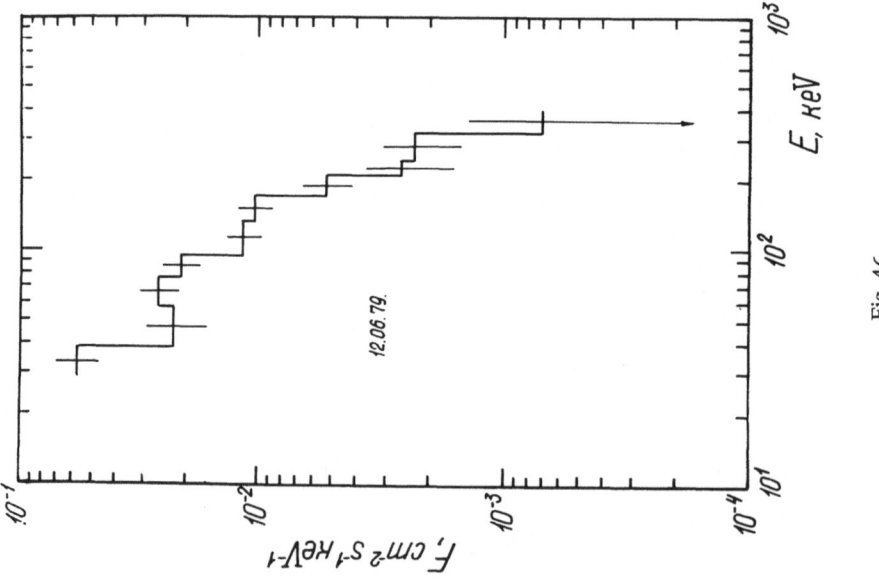

Fig. 46

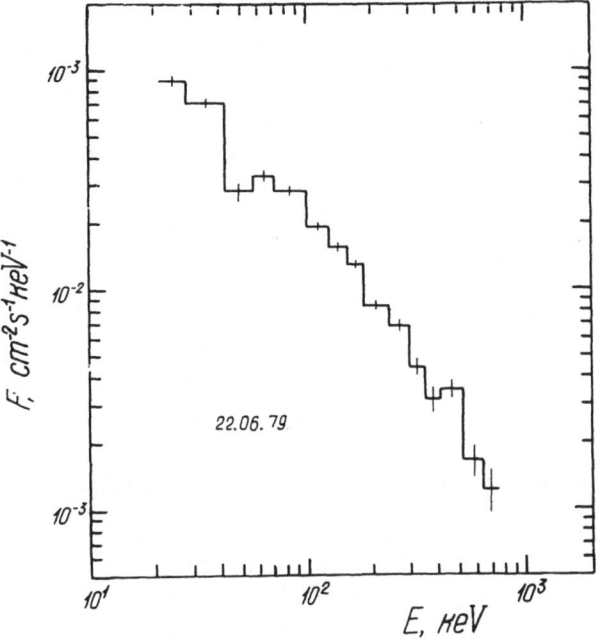

Fig. 48

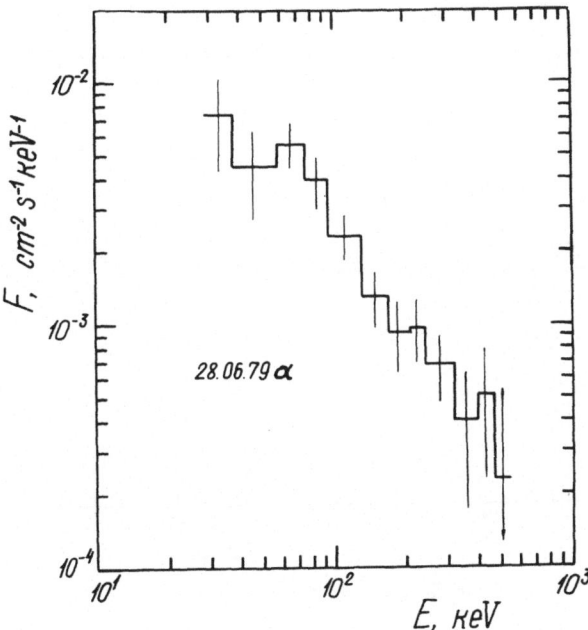

Fig. 49

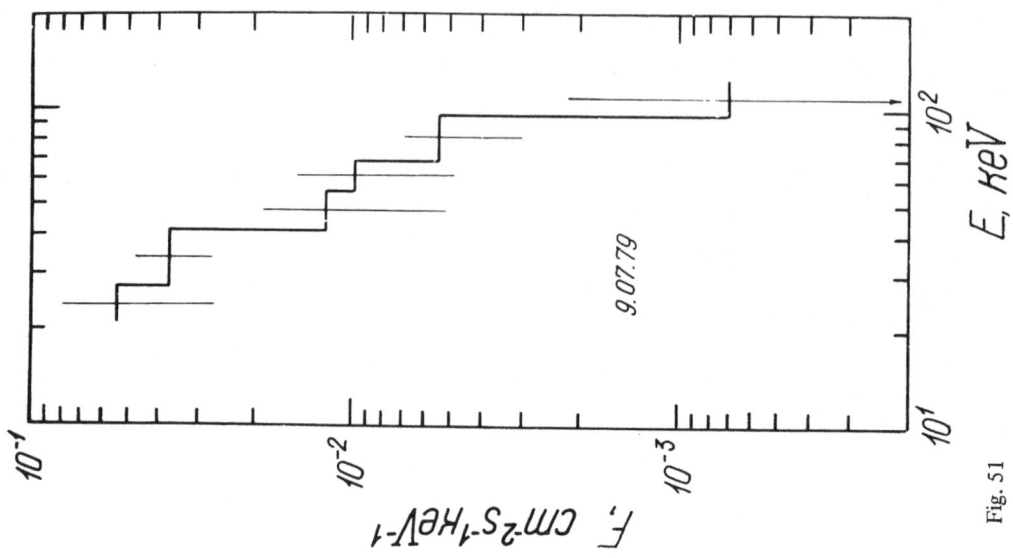

Fig. 51

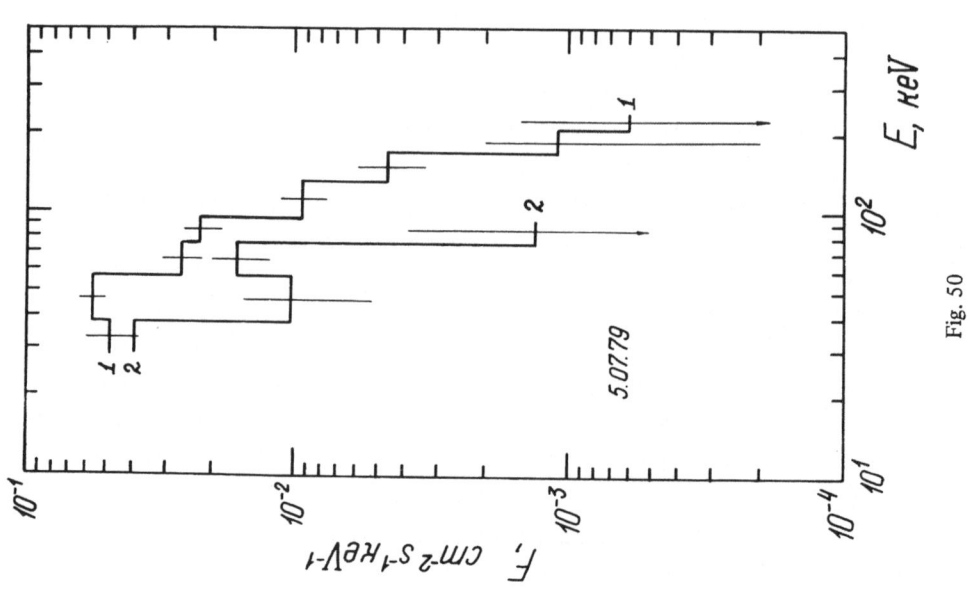

Fig. 50

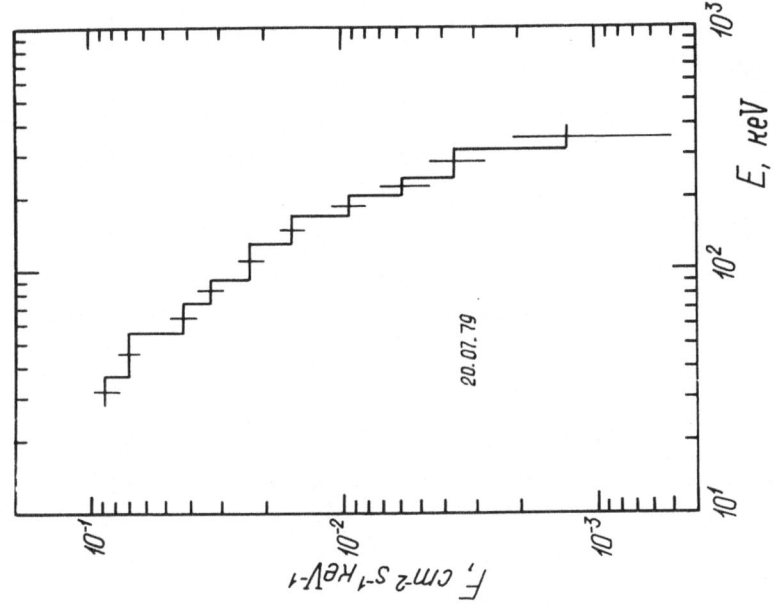

Fig. 53

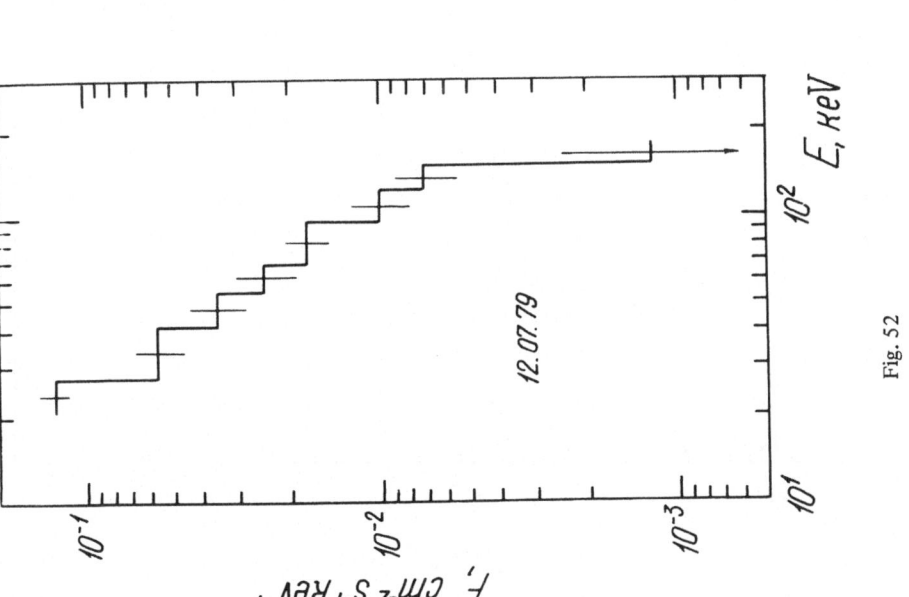

Fig. 52

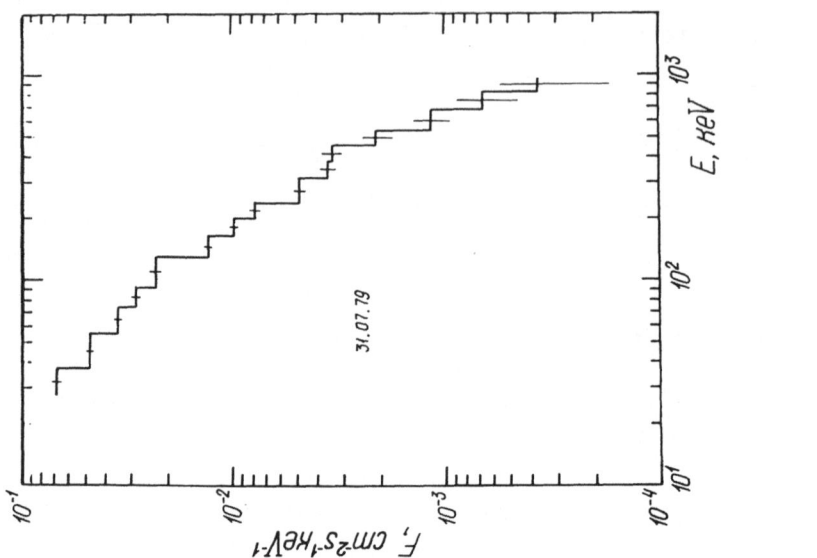

Fig. 55

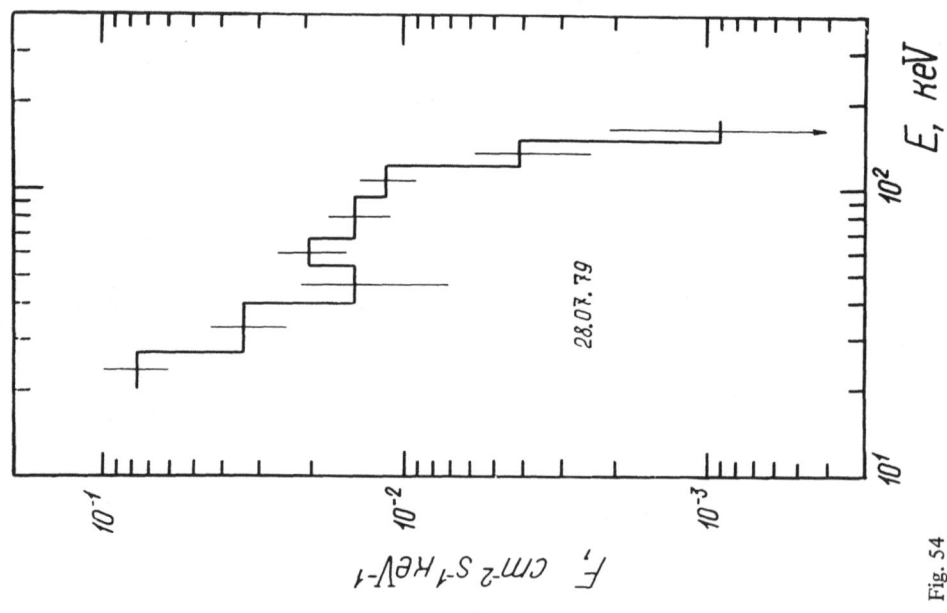

Fig. 54

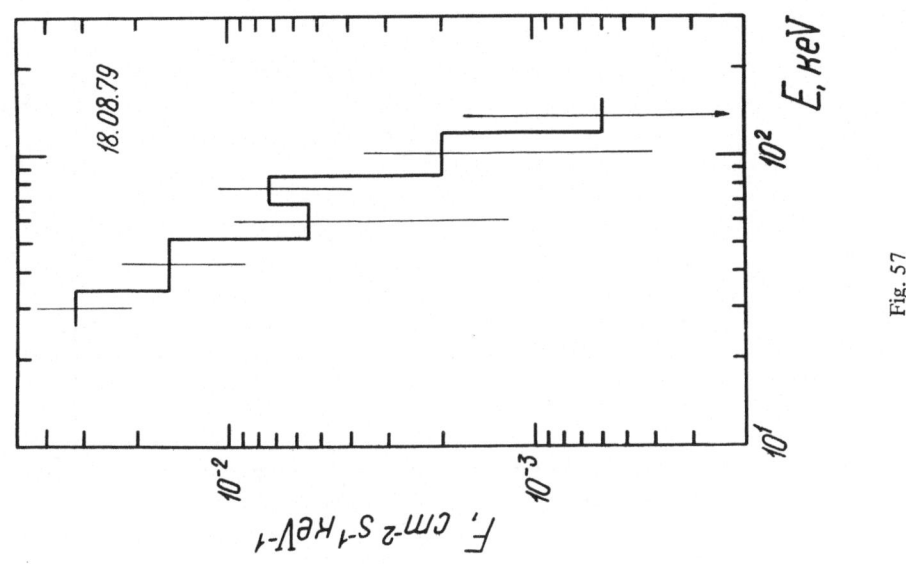

Fig. 57

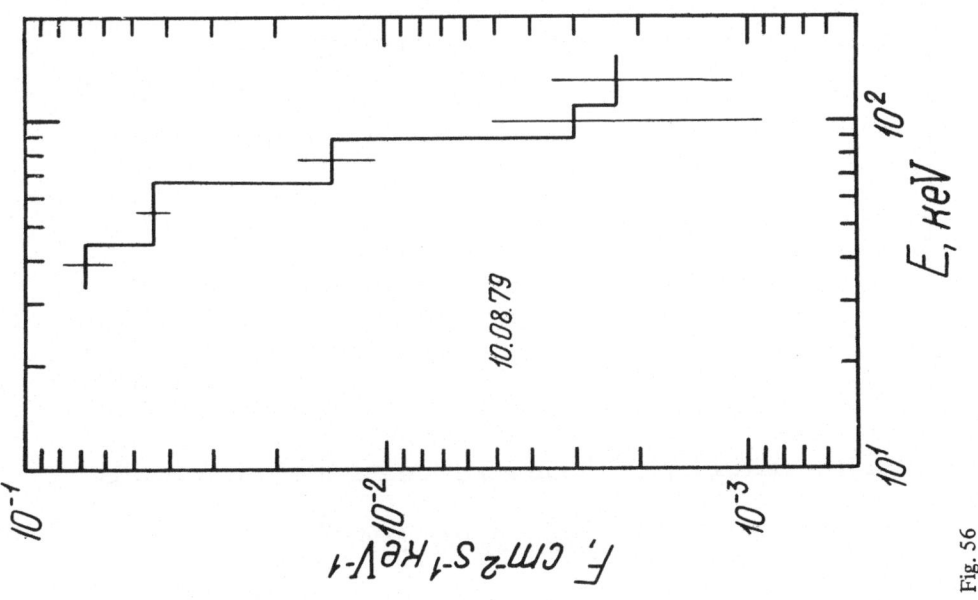

Fig. 56

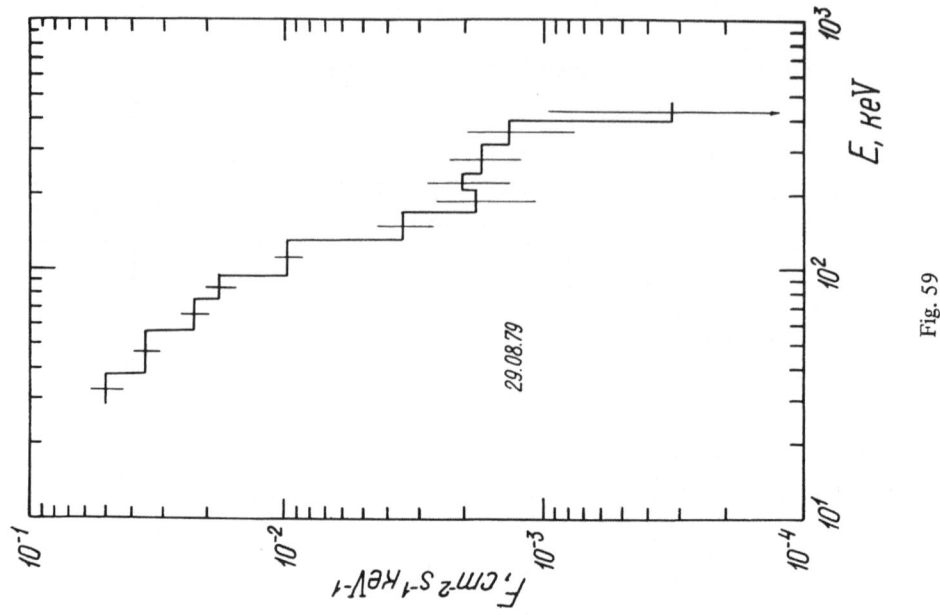

Fig. 59

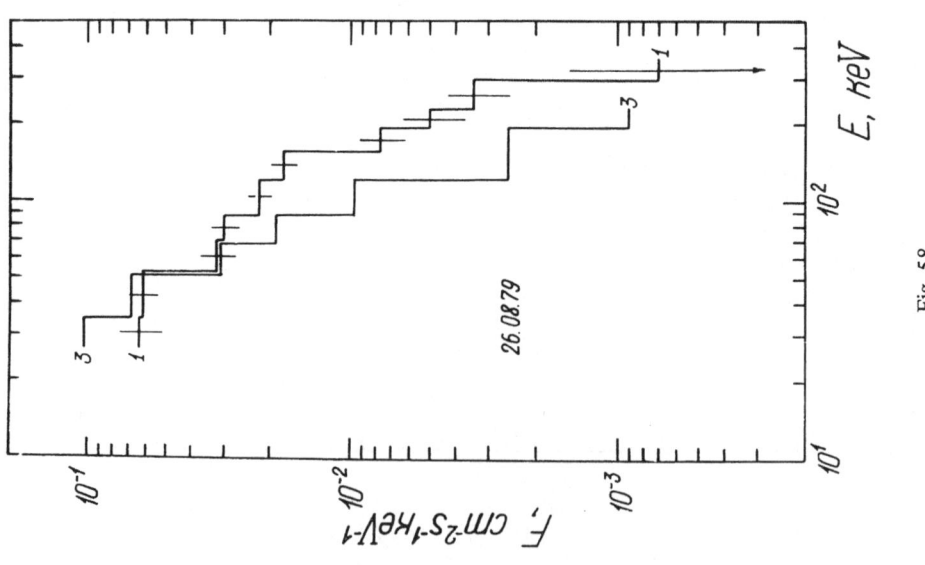

Fig. 58

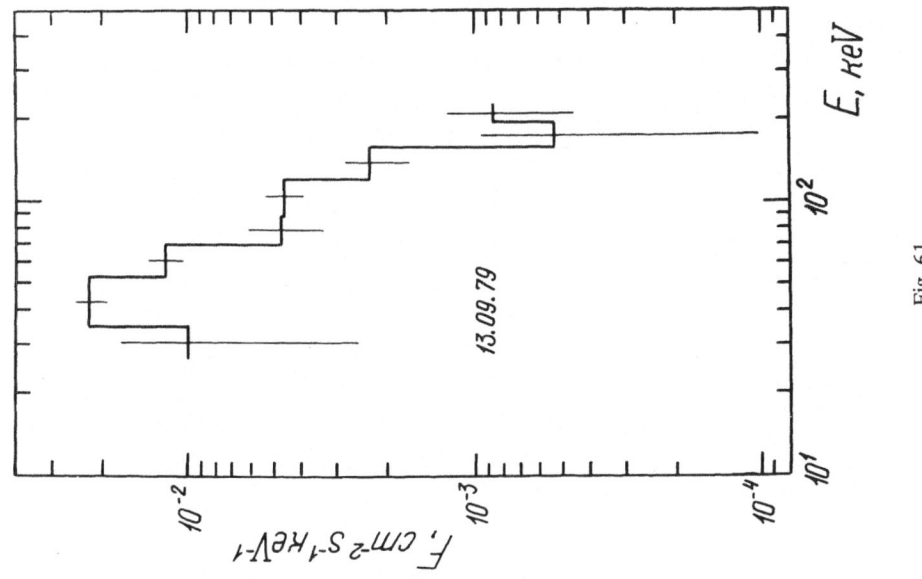

Fig. 61

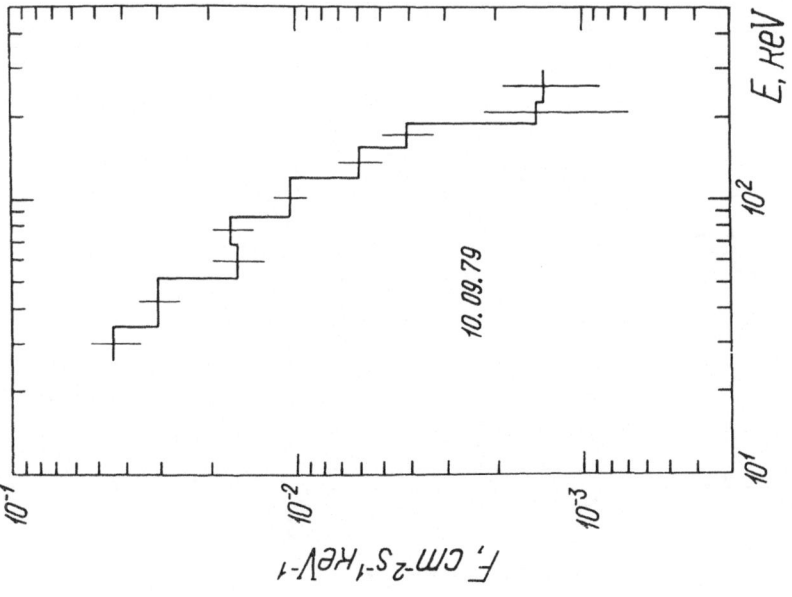

Fig. 60

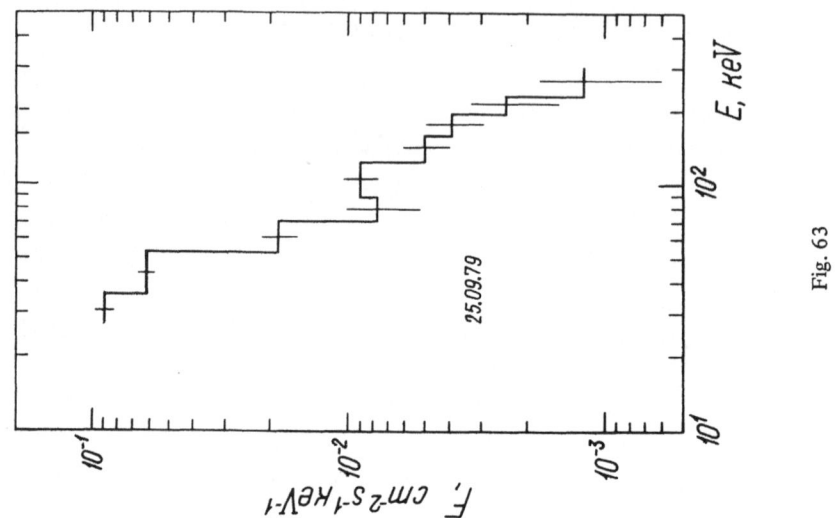

Fig. 63

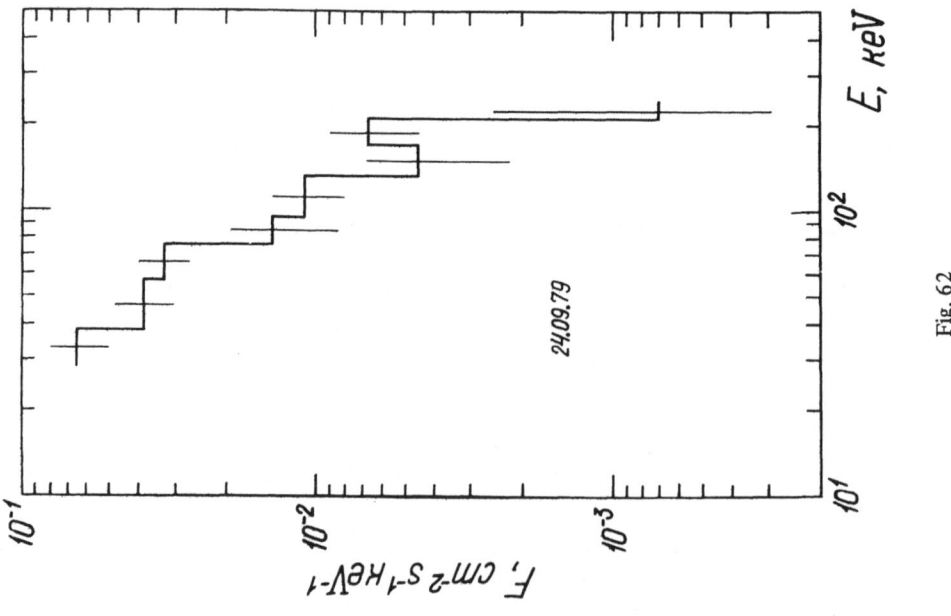

Fig. 62

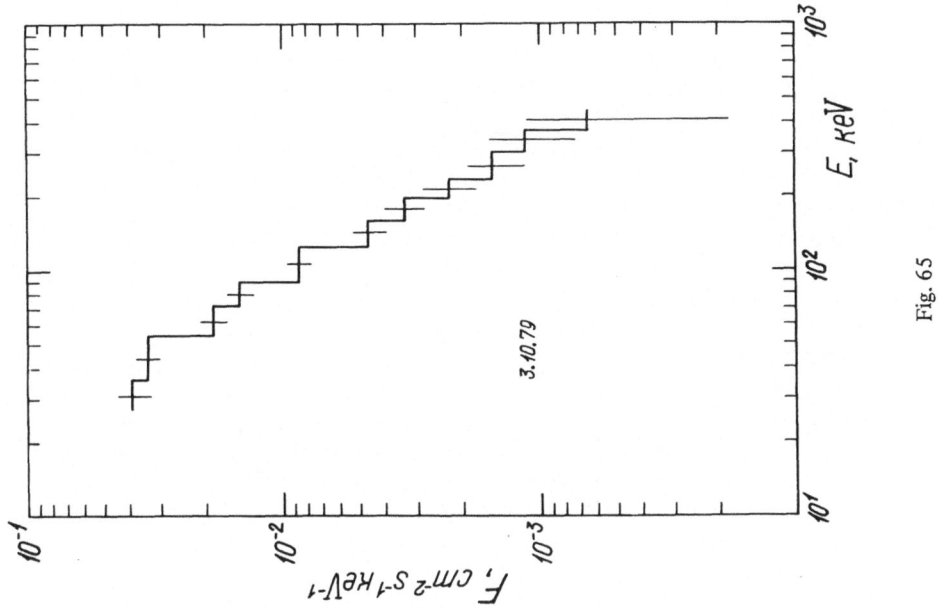

Fig. 65

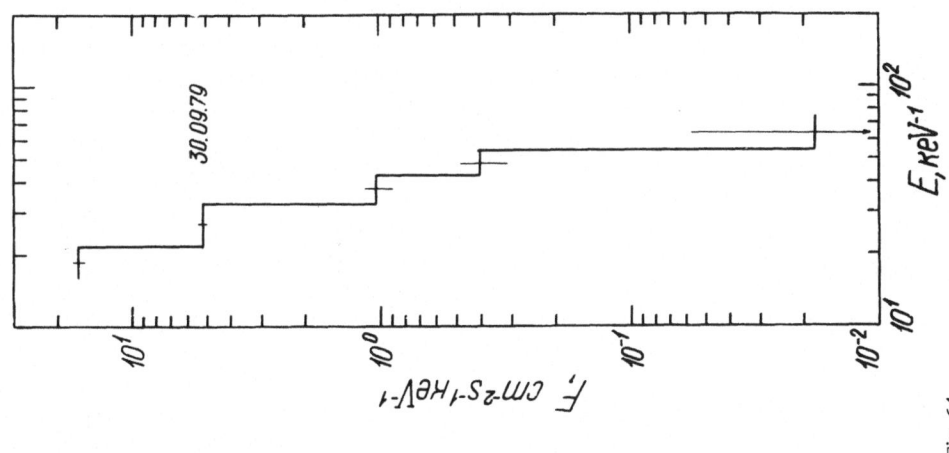

Fig. 64

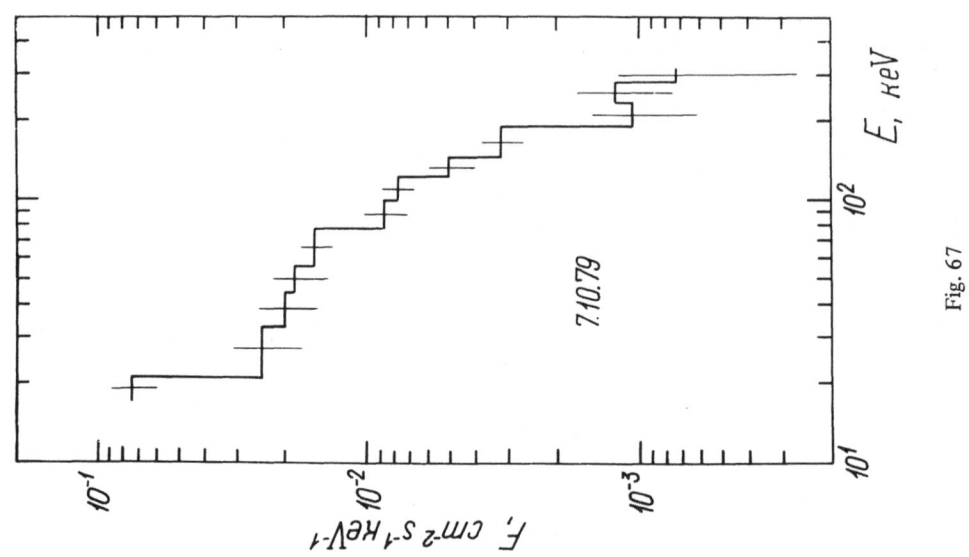

Fig. 67

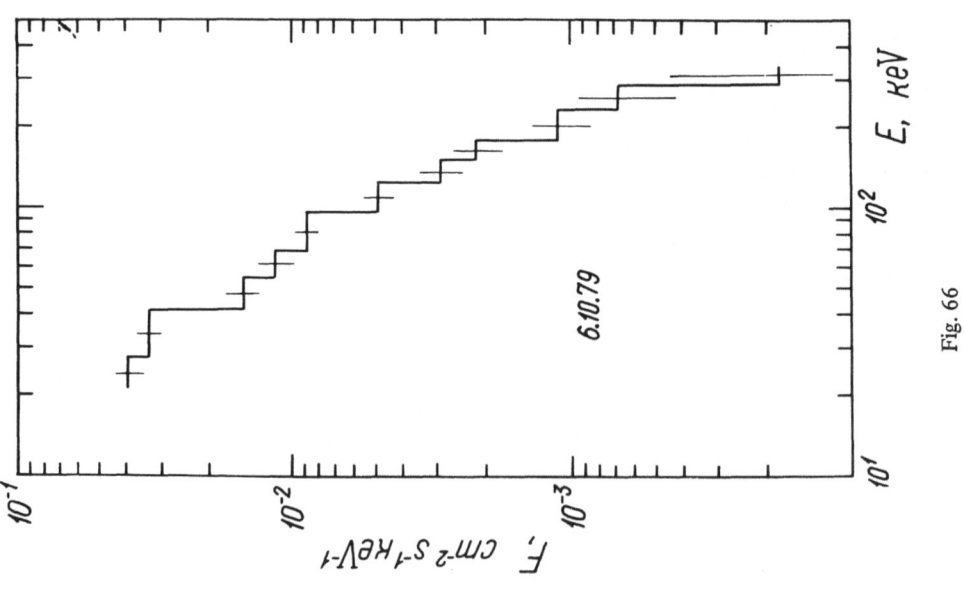

Fig. 66

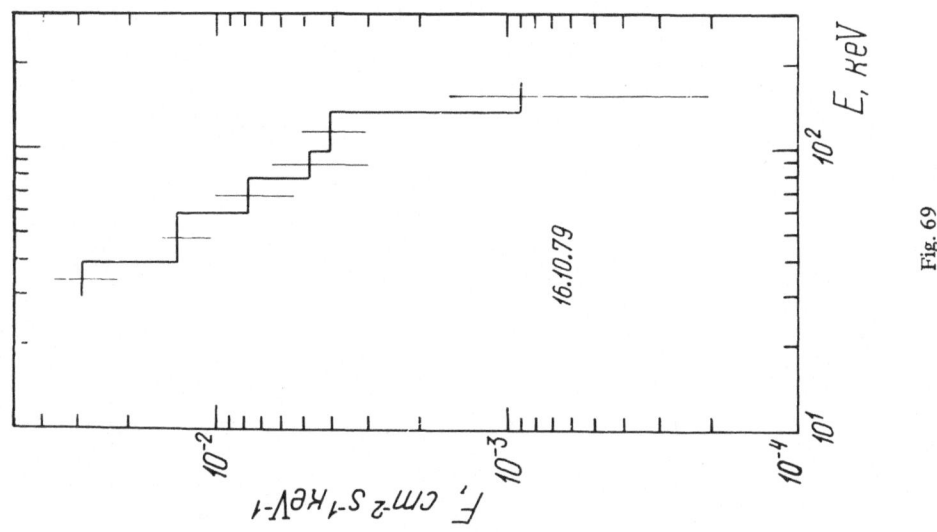

Fig. 69

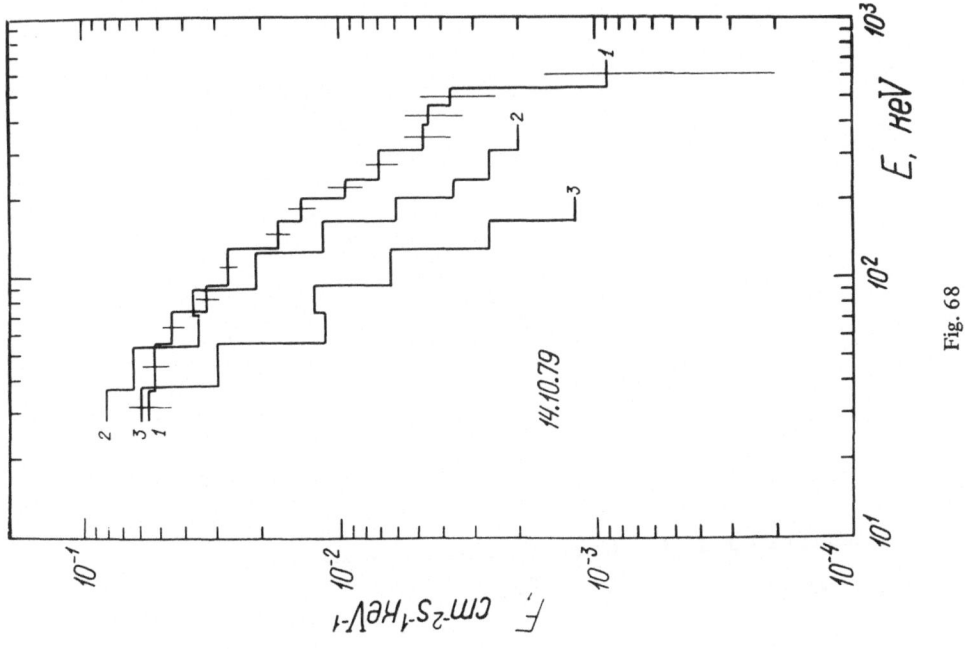

Fig. 68

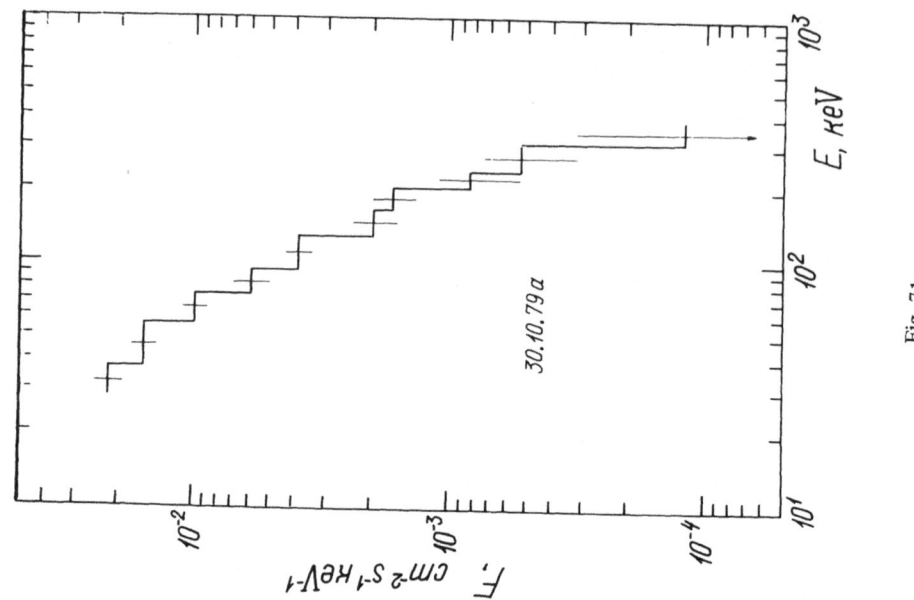

Fig. 71

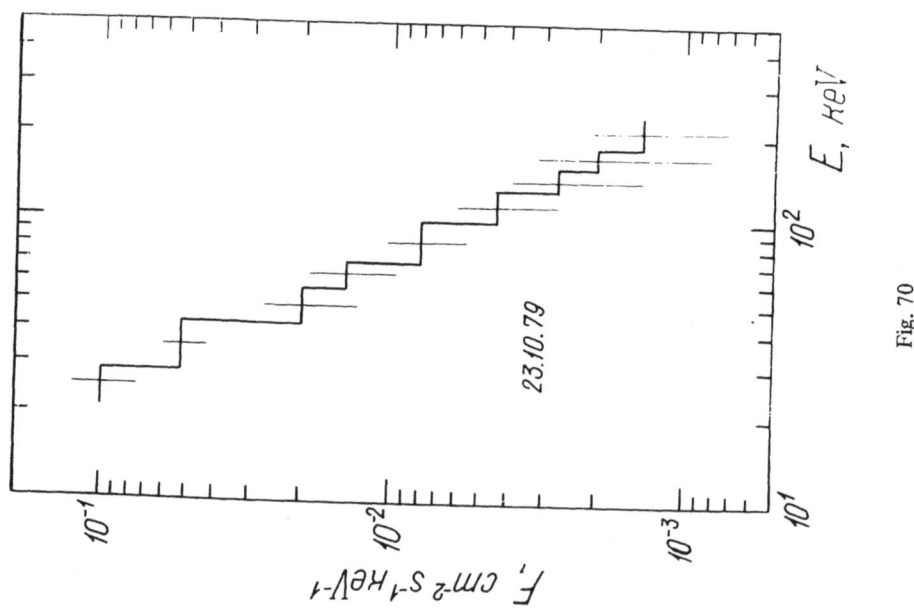

Fig. 70

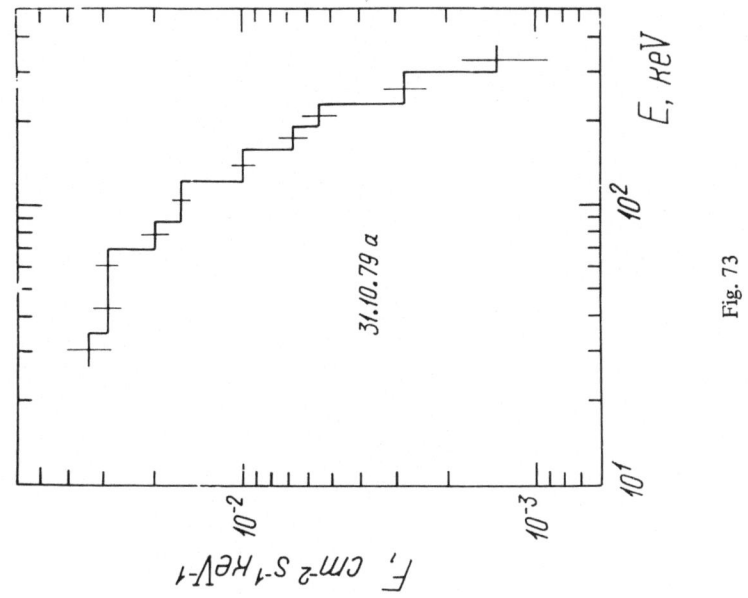

Fig. 73

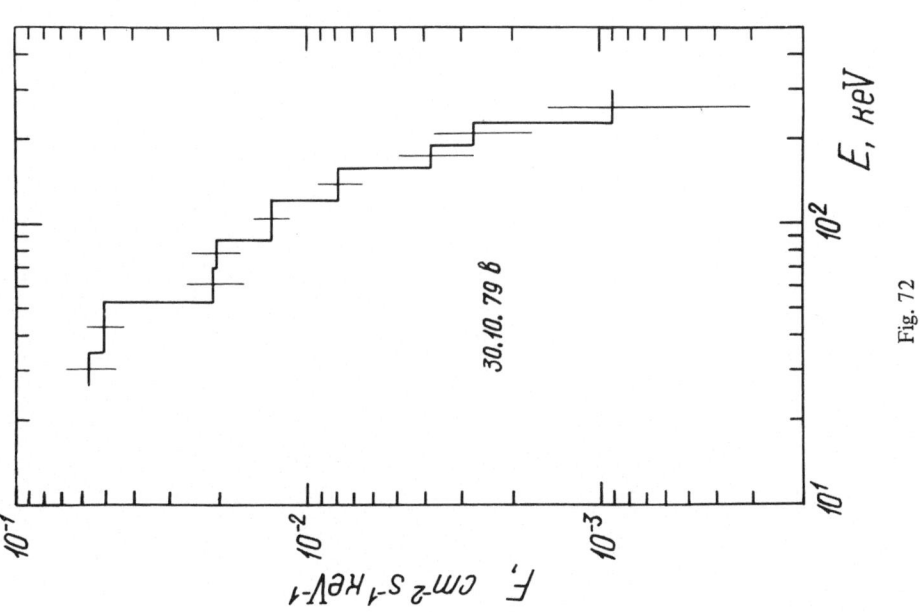

Fig. 72

CATALOG OF COSMIC GAMMA-RAY BURSTS FROM THE KONUS EXPERIMENT DATA

Part IV

E. P. MAZETS, S. V. GOLENETSKII, V. N. ILYINSKII, V. N. PANOV,
R. L. APTEKAR, Yu. A. GURYAN, M. P. PROSKURA, I. A. SOKOLOV,
Z. Ya. SOKOLOVA, and T. V. KHARITONOVA

A. F. Ioffe Physical-Technical Institute, U.S.S.R. Academy of Sciences, Leningrad, U.S.S.R.

and

A. V. DYATCHKOV and N. G. KHAVENSON

Institute of Cosmic Research, U.S.S.R. Academy of Sciences, Moscow, U.S.S.R.

(Received 18 May, 1981)

Abstract. The fourth and concluding part of the Catalog contains information on 20 gamma-ray bursts detected in the KONUS experiment on the Venera space probes from November 1979 up to the beginning of February 1980. The time profiles, energy spectra and source localization data are presented. A review of the major scientific results obtained is given.

1. Observations

In the KONUS experiment, from November 1979 up to the beginning of February 1980 (total exposure time 45 days), 20 gamma bursts were detected. The observational methods and instrumentation employed were described elsewhere (Mazets and Golenetskii, 1981a; Mazets *et al.*, 1979a). Information on the gamma-burst observations from September 1978 to October 1979 has been given in the preceding parts of the Catalog (Mazets *et al.*, 1981b, c).

A list of the bursts and of their parameters is given in Table I. The first column contains the dates of events, which also serve as event notation. The detection times T_0 for Venera-11 and Venera-12 are given in columns 2 and 3, respectively. Column 4 gives the burst intensity S which is the energy flux integrated over the burst duration. Columns 5 and 6 contain the center position and radius of the triangulation ring on the celestial sphere, obtained from the burst arrival times on the two spacecraft. For unambiguously localized sources, columns 7 and 8 give the mean source location on the celestial sphere in equatorial and galactic coordinates; the coordinates of the quadrangular localization box are presented in column 9. The equatorial coordinates are given for the epoch 1950.0 In the period in question, an unambiguous localization of 5 burst sources has been achieved.

Astrophysics and Space Science **80** (1981) 119–143. 0004–640X/81/0801–0119$03.75.

TABLE I

Gamma burst observations in the KONUS experiment, November 1979 to February 1980

Date of event	T_0, h.m.s. UT V-11	V-12	$S(E\gamma > 30\,\text{keV})$ erg cm^{-2} $\times 10^{-6}$	Triangulation ring — Ring center α δ	Ring radius	Mean source position α δ	l'' b''	Box coordinates α δ			
(1)	(2)	(3)	(4)	(5)	(6)	(7)	(8)	(9)			
1.11.79	No data	17.36.27	240								
5.11.79a	04.07.40	04.10.23	3.5	194.971 / −6.946	63.067 ± 0.045	251.634 / 23.346	42.1 / 36.7	253.021 / 20.000	250.098 / 26.772	252.912 / 20.021	249.973 / 26.801
5.11.79b	13.43.43	13.38.45	13	15.458 / 7.178	33.895 ± 0.029	342.083 / 0.721	71.5 / −49.7	342.659 / −1.754	341.612 / 3.243	342.712 / −1.731	341.666 / 3.266
9.11.79	No data	07.56.32	6.8			130.4 / −33.6	254.8 / 5.5	126.7 / −33.0	129.7 / −37.2	131.1 / −29.8	134.1 / −34.0
11.11.79	No data	11.18.27	10								
13.11.79	No data	00.03.34	9.0								
15.12.79	06.07.13	06.02.19	10	69.006 / 24.824	29.840 ± 0.170	50.88 / 51.24	145.4 / −4.3	54.52 / 52.77	47.37 / 49.68	54.86 / 52.49	47.73 / 49.44
18.12.79	16.44.29	No data	0.47								
20.12.79	No data	17.30.34	100								
22.12.79	12.42.51	12.40.57	1.1	79.429 / 26.060	69.854 ± 0.046						
26.12.79	No data	17.03.40	0.70								
29.12.79	15.38.58	Not detected	3.2								
30.12.79	09.12.57	09.13.36	350	270.572 / −26.556	82.850 ± 0.091	258.13 / 55.61	82.8 / 35.8	264.20 / 56.20	252.02 / 54.83	264.43 / 56.04	252.29 / 54.69
2.01.80	Not detected	15.04.23	0.47								
5.01.80	No data	10.49.04	37								
12.01.80	00.59.00	No data	2.4								
16.01.80	15.51.23	No data	21								

2. Burst Temporal Structure

The time profiles of the gamma bursts detected are shown in Figures 1–21. The graphs show the count rate in the energy window 50–150 keV vs. time with 0.25 s resolution (after $T - T_0 = 34$ s, with 1 s resolution). Separate data points before the continuous histogram represent the prehistory of the gamma burst recorded with the same detector. The horizontal dashed line shows the mean background level measured before and after a gamma burst. For the event 5 November, 1979b which has an intensive initial phase, a record of the burst initial part with 1/64 s resolution is also presented (Figure 4). For the event 26 December, 1979, which belongs to the separate class of short bursts (see below) only a time profile recording with 1/64 s resolution is given.

3. Energy Spectra

The differential energy spectra of gamma bursts in the range 30 keV–2 MeV are presented in Figures 22–37. Several weakest bursts are not presented here, their spectral information being not statistically reliable.

Eight energy spectra are recorded in the KONUS experiment for each burst during eight consecutive time intervals of 4 s each. Only spectra summarized over all burst duration are presented in most of the graphs. For the events 9 and 11 November 1979, revealing a strong temporal evolution of the spectrum, we present spectra obtained in several 4 s time intervals after T_0. The corresponding curves in the graphs are labelled accordingly.

4. Major Results of the Experiment

The KONUS experiment was carried out for one-and-a-half years. In the 385 days of observation, from September 1978 to February 1980, 143 gamma bursts have been detected altogether. First of all, this result provided support for the assumption that the gamma bursts are not uniquely rare events. At the sensitivity level achieved, one gamma burst is observed every two to three days. The large volume of high accuracy, detailed information accumulated by two identical instruments in the same space experiment, has substantially broadened our knowledge on the gamma bursts. Novel results of a fundamental nature have been obtained.

Despite the extreme diversity of the gamma burst observed, several typical time profile structures of the events have been established to exist which probably reflects important features of the gamma-ray generation processes in the sources. Short bursts make up a separate class, their distinctive features being so specific as to make one assume the short and the longer bursts to be of different origin (Mazets and Golenetskii, 1981a).

Individual differences in the energy spectra are pronounced to a much lesser extent than those in the burst time profiles. Radiation from an optically thin, hot plasma is the best fit for the energy spectrum shape observed (Mazets *et al.*, 1980a). The temperatures for different bursts range from 10^8 to 10^{10} K. The spectra are frequently seen to evolve in time, the temperature quite often becoming correlated with gamma-ray intensity.

The spectra of many gamma bursts have revealed emission and absorption lines (Mazets *et al.*, 1980a, 1981d). The similarity of the spectral features observed in different events lends strong support to the proposed explanation of the nature of these lines involving processes which can occur only in a hot plasma near the surface of neutron stars with a strong magnetic field. The emission lines represent annihilation radiation from the electrons and positrons undergoing strong gravitational redshift, $z \approx 0.2$–0.3. The absorption lines are due to resonant magnetic bremsstrahlung absorption of burst radiation at the electron cyclotron frequency at fields of $(2-7) \times 10^{12}$ G. Evolution is a typical feature of the lines. They are observed, as a rule, at the initial stage of a burst.

The source of four successive bursts on 5 and 6 March, 4 and 24 April 1979 has been established to be a flaring X-ray pulsar in Dorado (Mazets *et al.*, 1979b, c; Golenetskii *et al.*, 1979). A constant-period pulsation of radiation in the tail of the strong burst on 5 March 1979, and the redshifted annihilation line in its spectrum observed at the initial stage imply that this source contains a strongly magnetized neutron star rotating with an 8.1 s period.

One more source of recurrent short bursts has been detected (Mazets *et al.*, 1979d). The recurrence of the short bursts supports their being different from longer events.

The information obtained has been statistically analyzed. Gamma burst appearance frequency vs. observed total burst energy S and power P plots, as well as the angular distributions of the sources on the celestial sphere have been constructed and studied. A strong correlation between burst intensity and duration was observed. These results not only provide a proof for the galactic origin of the gamma bursts but also permit one to estimate the parameters of the source spatial distribution in the Galaxy (Mazets *et al.*, 1980b).

On the whole, the results obtained indicate with certainty that the gamma burst sources contain neutron stars possessing a strong magnetic field. The most likely model for the burst source is a binary made up of a strongly magnetized, slowed-down neutron star and a normal star of low mass and luminosity in which nonstationary accretion occurs. The short bursts are apparently due to explosive thermonuclear burning of matter falling onto the neutron star's surface.

References

Golenetskii, S. V., Mazets, E. P., Ilyinskii, V. N., and Guryan, Yu. A.: 1979, *Astron. Zh.-Pisma* **5**, 636.
Mazets, E. P., Golenetskii, S. V., Ilyinskii, V. N., Panov, V. N., Aptekar, R. L., Guryan, Yu. A., Sokolov, I. A., Sokolova Z. Ya. and Kharitonova, T. V.: 1979a, *Kosmitcheskiye Issledovaniya* **17**, 812.

Mazets, E. P., Golenetskii, S. V., Ilyinskii, V. N., Panov, V. N., Aptekar, R. L., Guryan, Yu. A., Sokolov, I. A., Sokolova, Z. Ya. and Kharitonova, T. V.: 1979b, *Astron. Zh.-Pisma* **5**, 307.

Mazets, E. P., Golenetskii, S. V., Ilyinskii, V. N., Aptekar, R. L., and Guryan, Yu. A.: 1979c, *Nature* **282**, 587.

Mazets, E. P., Golenetskii, S. V., and Guryan, Yu. A.: 1979d. *Astron. Zh.-Pisma* **5**, 641.

Mazets, E. P., Golenetskii, S. V., Aptekar, R. L., Guryan, Yu. A., and Ilyinskii, V. N.: 1980a, *Astron. Zh.-Pisma* **6**, 706.

Mazets, E. P., Golenetskii, S. V., Aptekar, R. L., Guryan, Yu. A., and Ilyinskii, V. N.: 1980b, *Astron. Zh.-Pisma* **6**, 609.

Mazets, E. P. and Golenetskii, S. V.: 1981a, *Astrophys. Space Sci.* **75**, 47.

Mazets, E. P., Golenetskii, S. V., Ilyinskii, V. N., Panov, V. N., Aptekar, R. L., Guryan, Yu. A., Proskura, M. P., Sokolov, I. A., Sokolova, Z. Ya., Kharitonova, T. V., Dyatchkov, A. V., and Khavenson, N. G.: 1981b, *Astrophys. Space Sci.* **80**, 3 (this issue).

Mazets, E. P., Golenetskii, S. V., Ilyinskii, V. N., Panov, V. N., Aptekar, R. L., Guryan, Yu. A., Proskura, M. P., Sokolov, I. A., Sokolova, Z. Ya., Kharitonova, T. V., Dyatchkov, A. V., and Khavenson, N. G.: 1981c, *Astrophys. Space Sci.* **80**, 85 (this issue).

Mazets, E. P., Golenetskii, S. V., Aptekar, R. L., Guryan, Yu. A., and Ilyinskii, V. N.: 1981d, *Nature* **290**, 378.

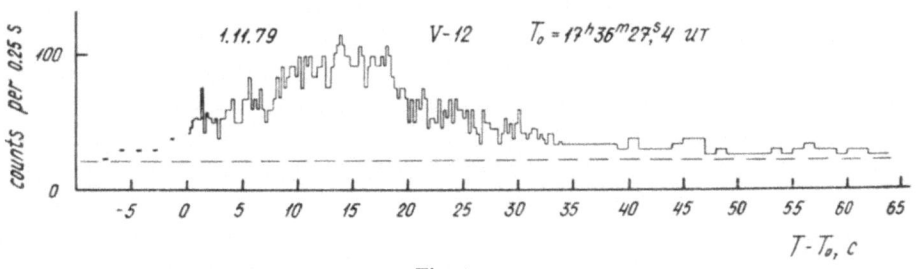

Fig. 1

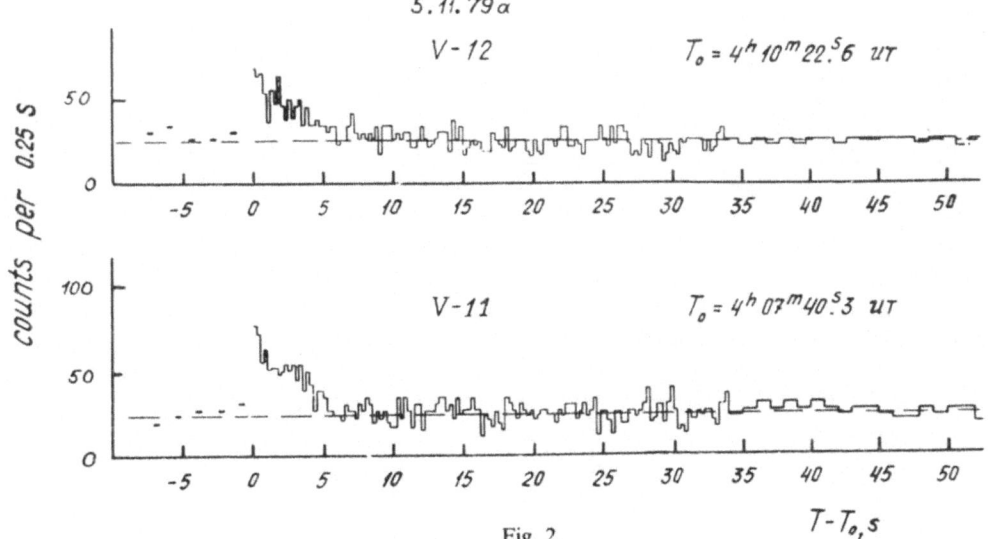

Fig. 2

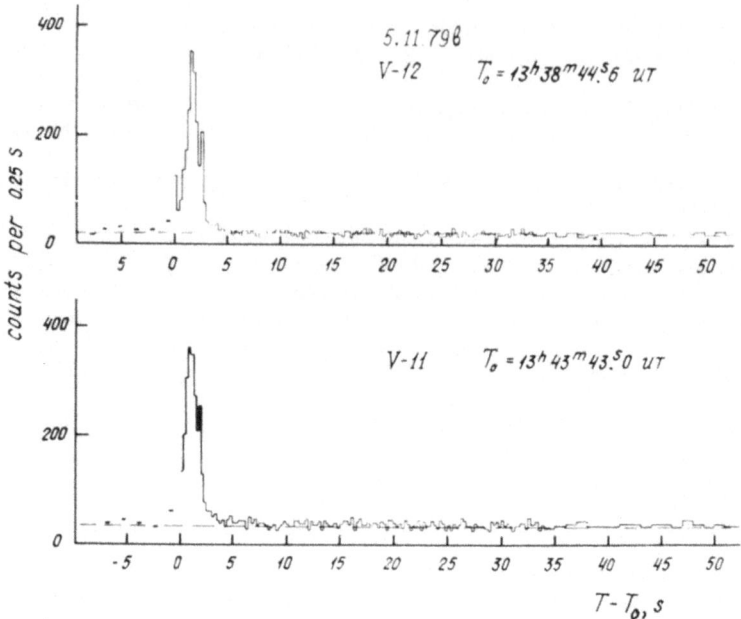

Fig. 3

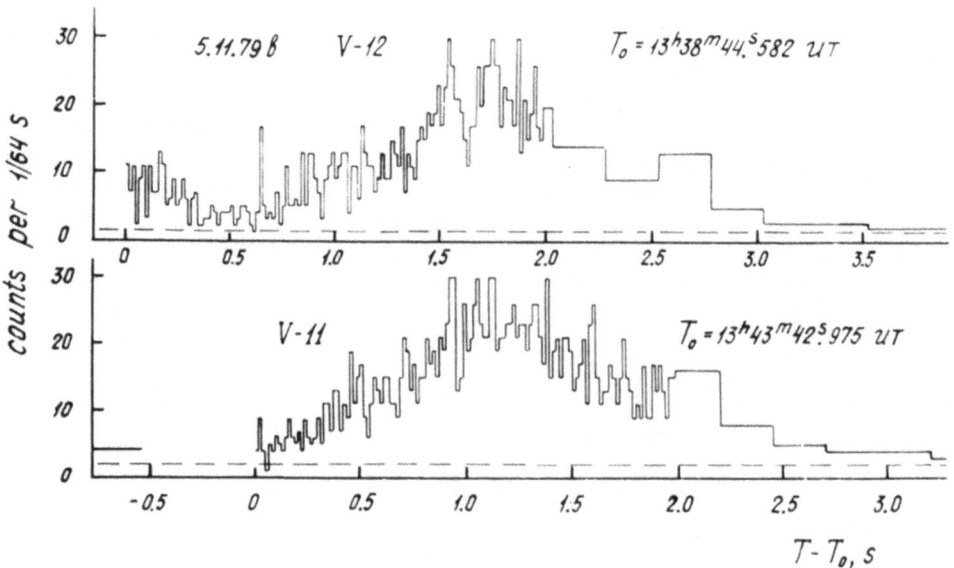

Fig. 4

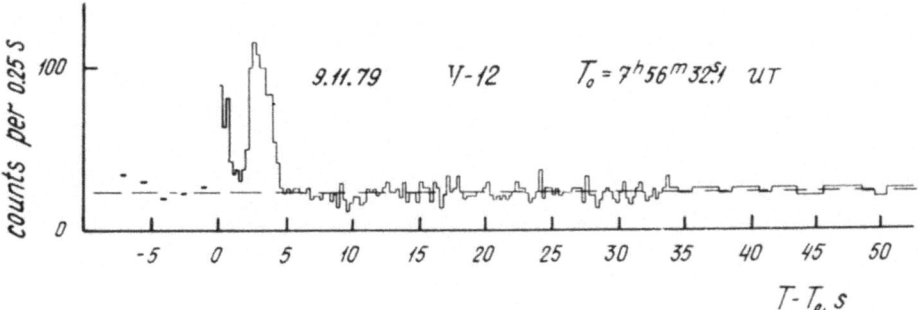

Fig. 5

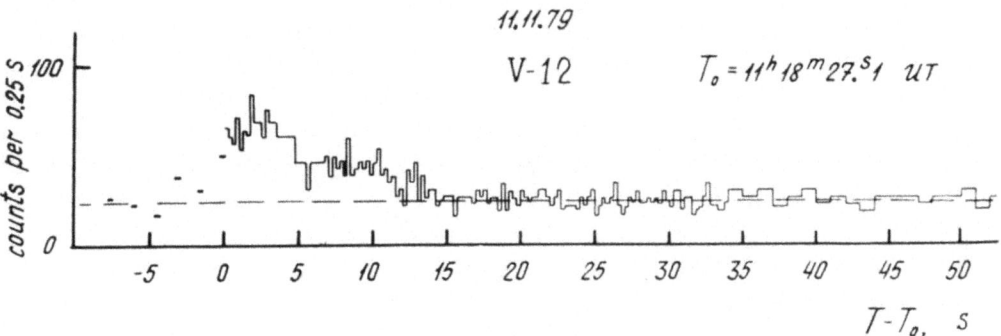

Fig. 6

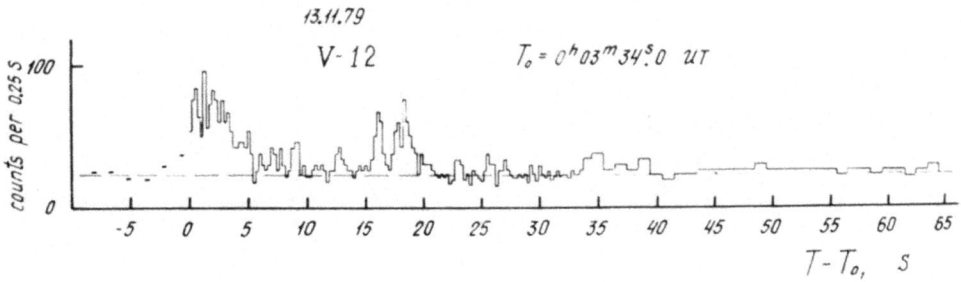

Fig. 7

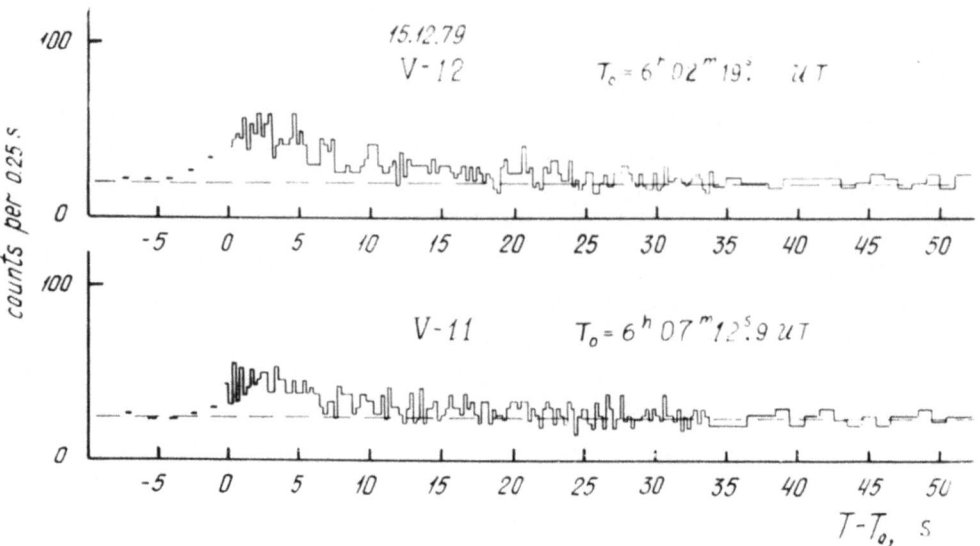

Fig. 8

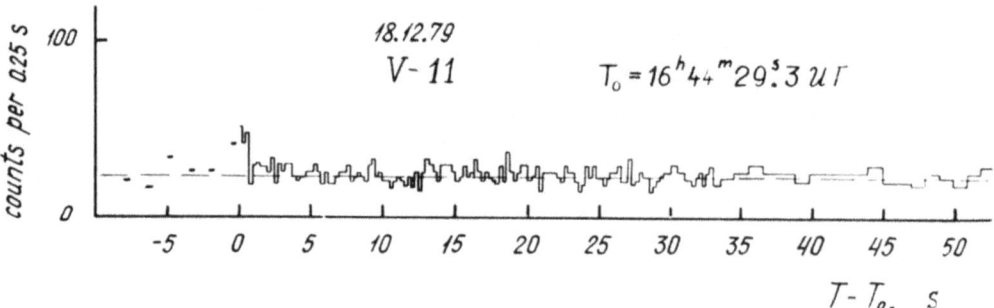

Fig. 9

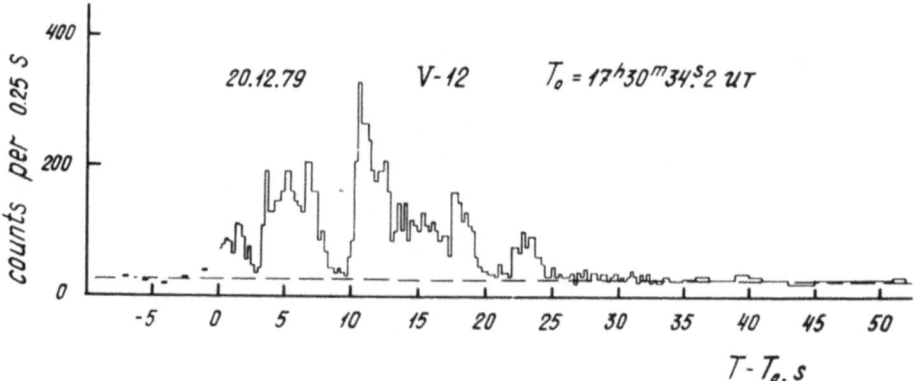

Fig. 10

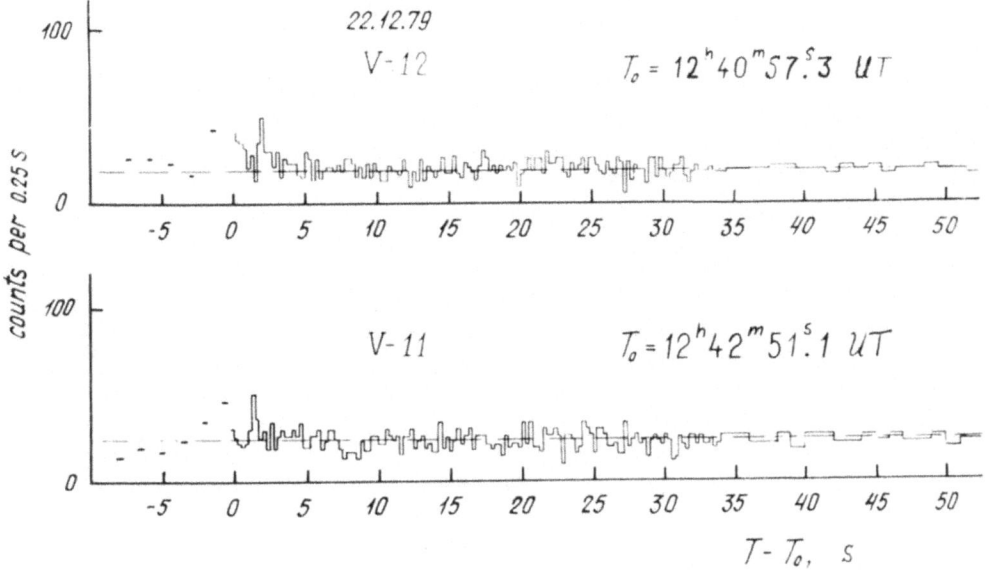

Fig. 11

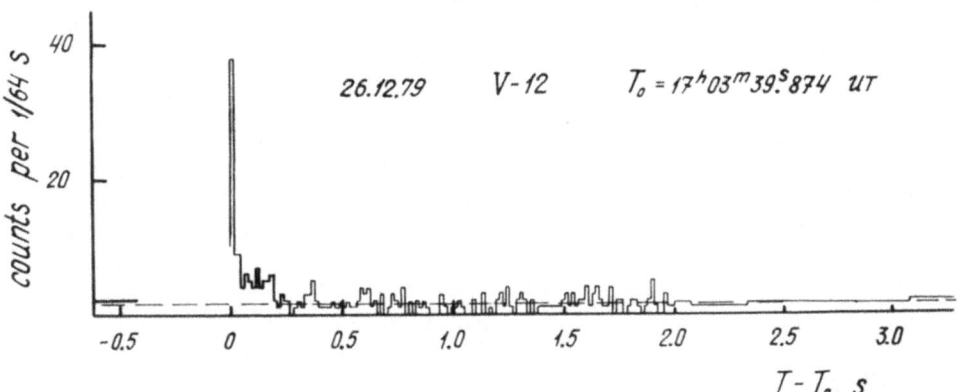

Fig. 12

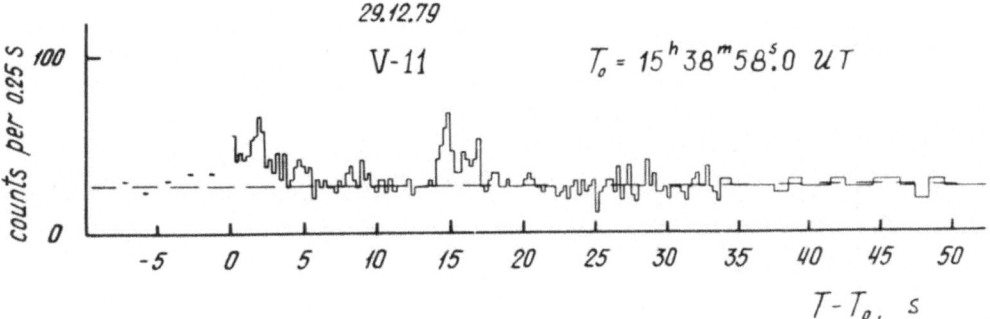

Fig. 13

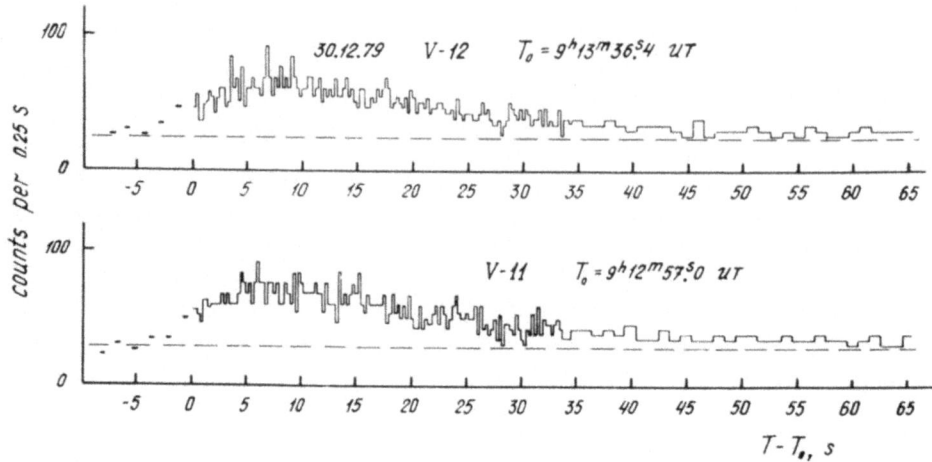

Fig. 14

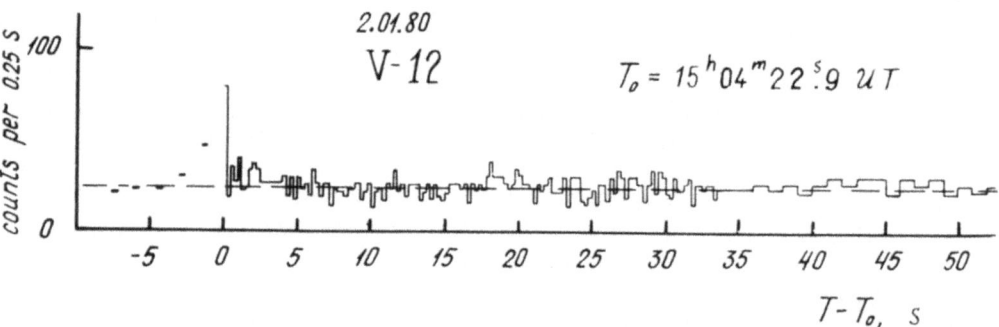

Fig. 15

Fig. 16

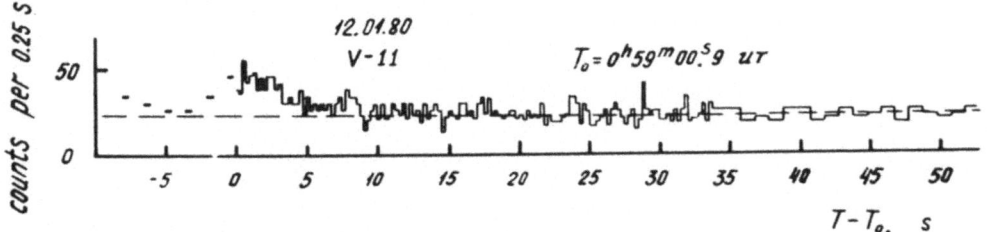

Fig. 17

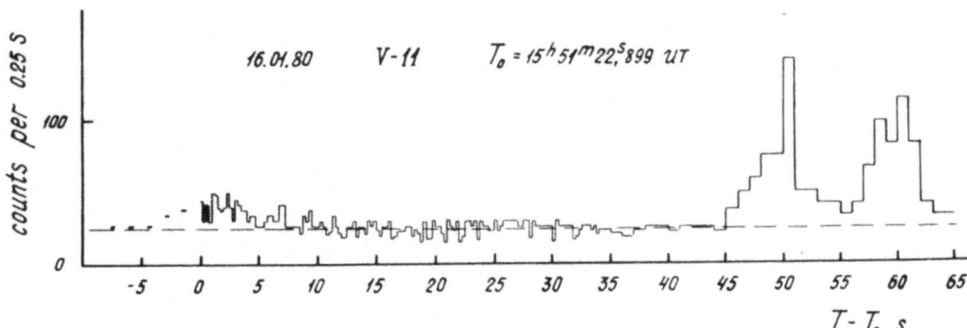

Fig. 18

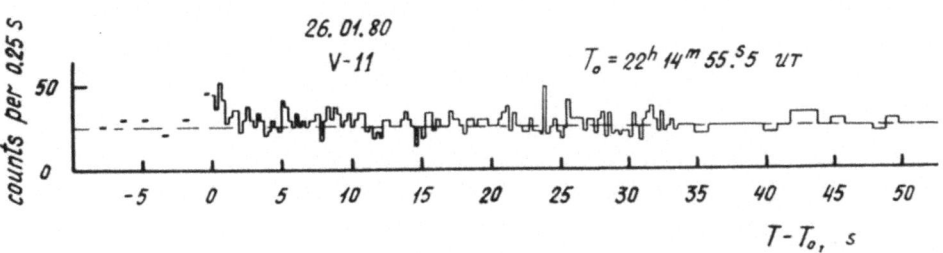

Fig. 19

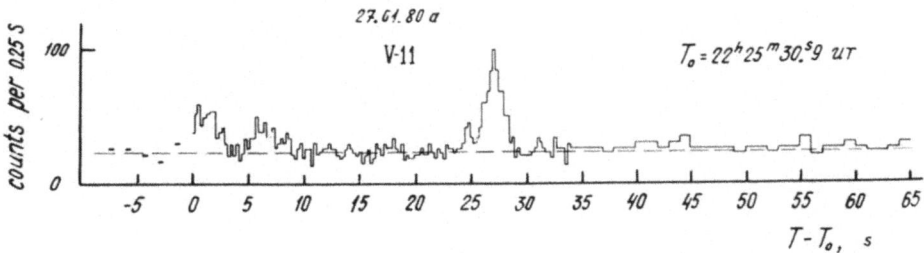

Fig. 20

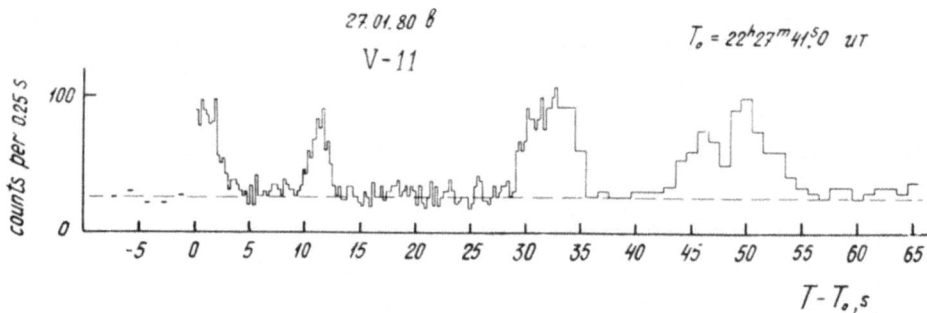

Fig. 21

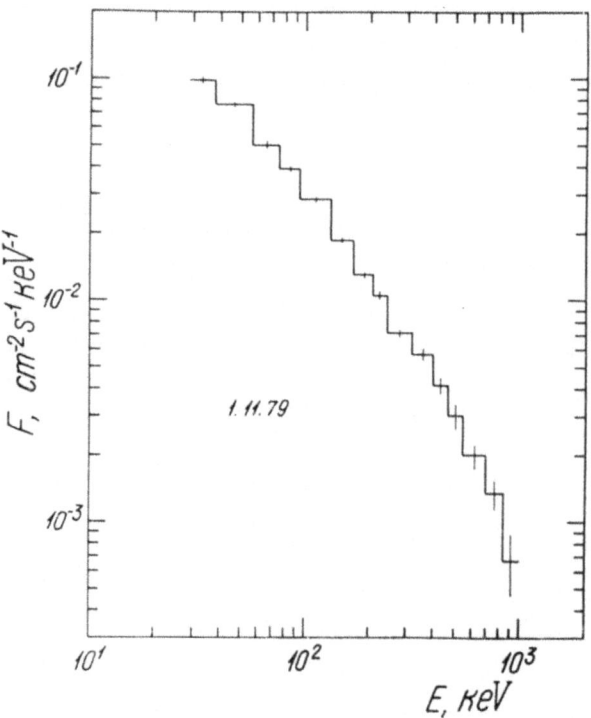

Fig. 22

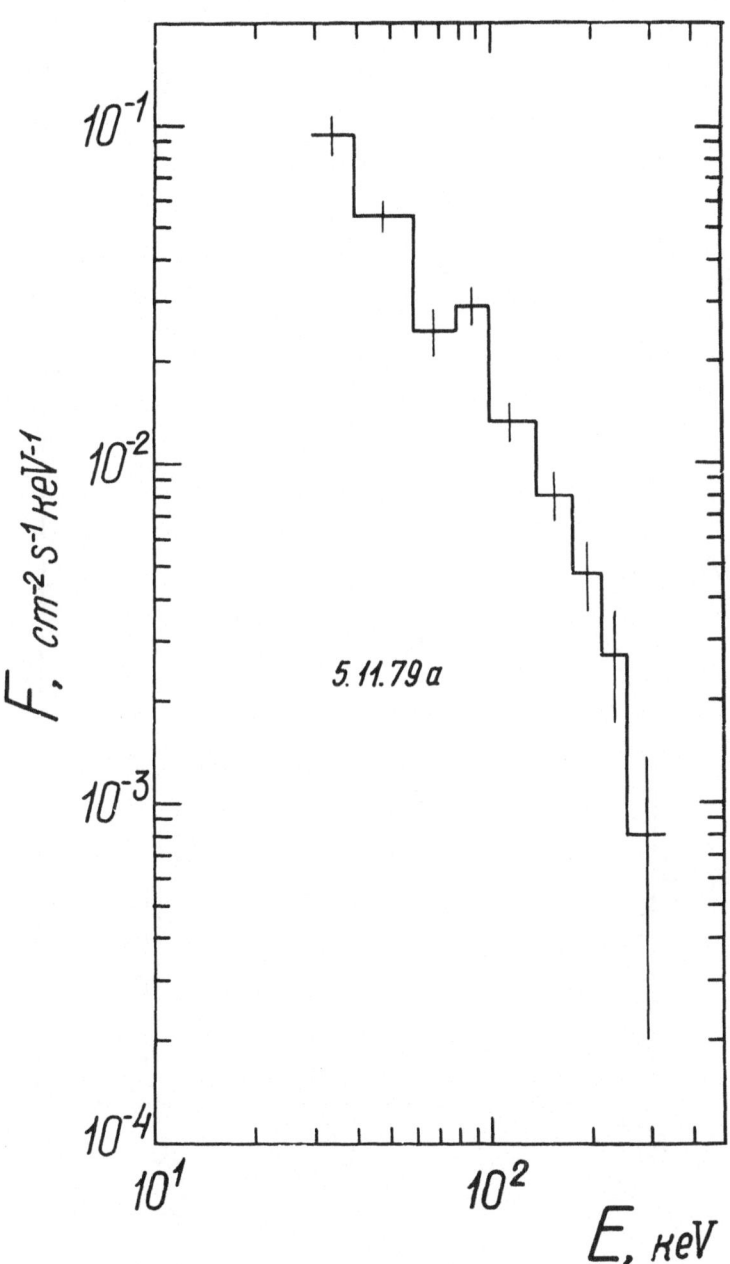

Fig. 23

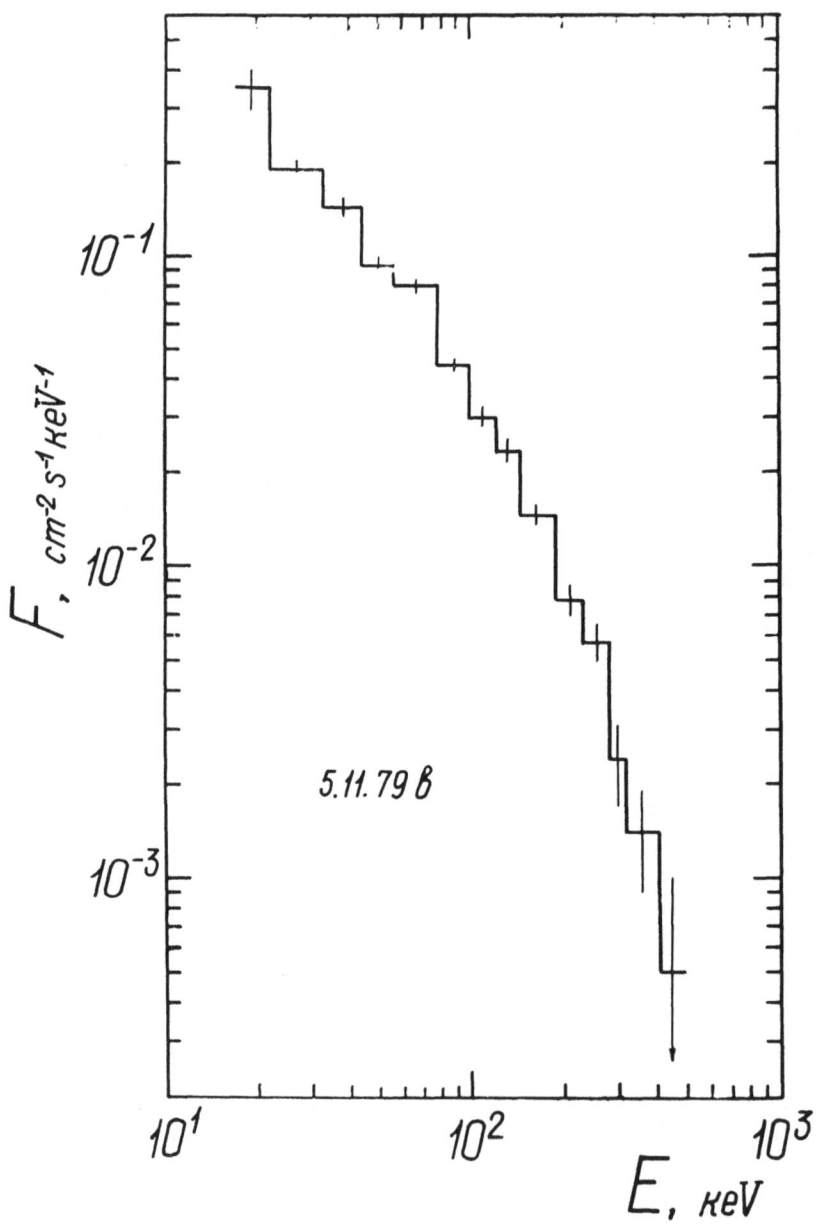

Fig. 24

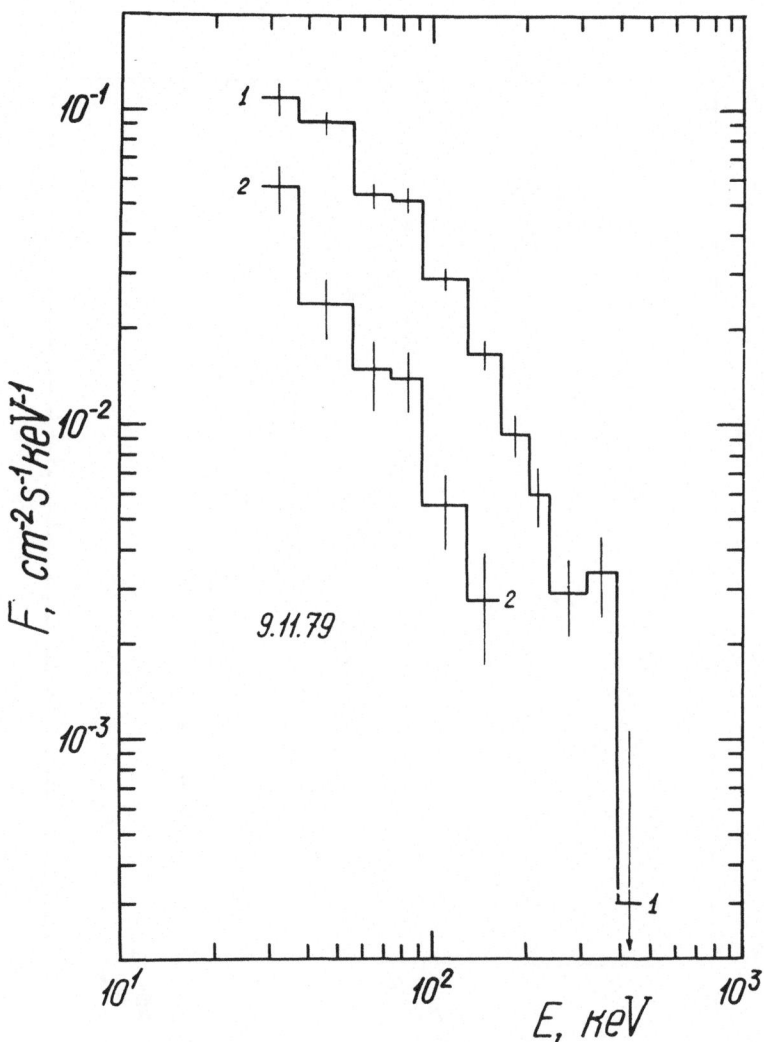

Fig. 25

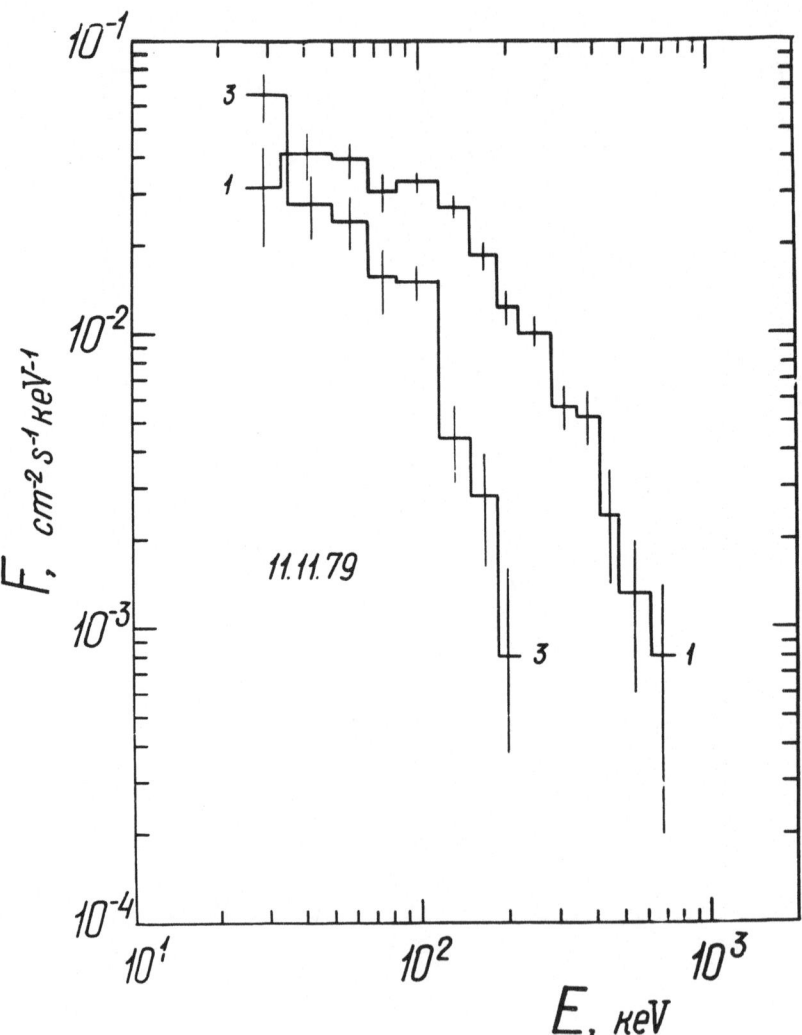

Fig. 26

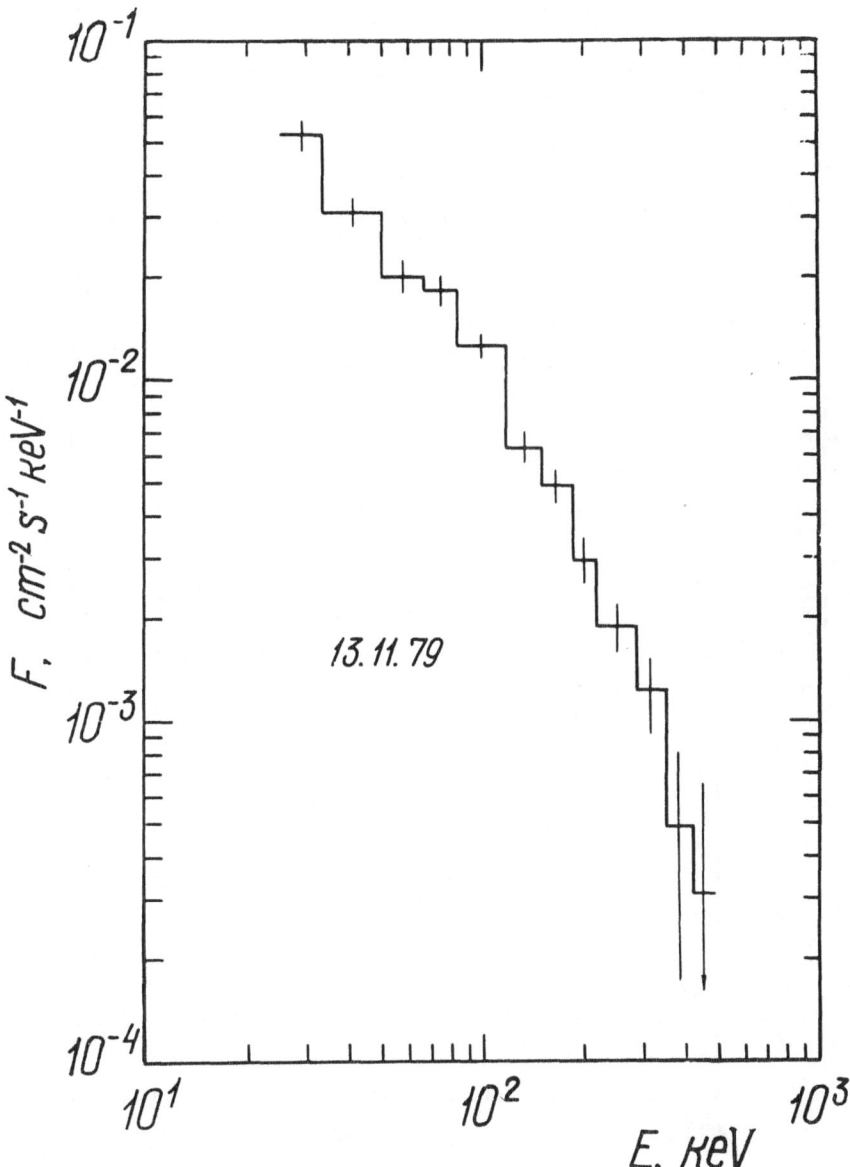

Fig. 27

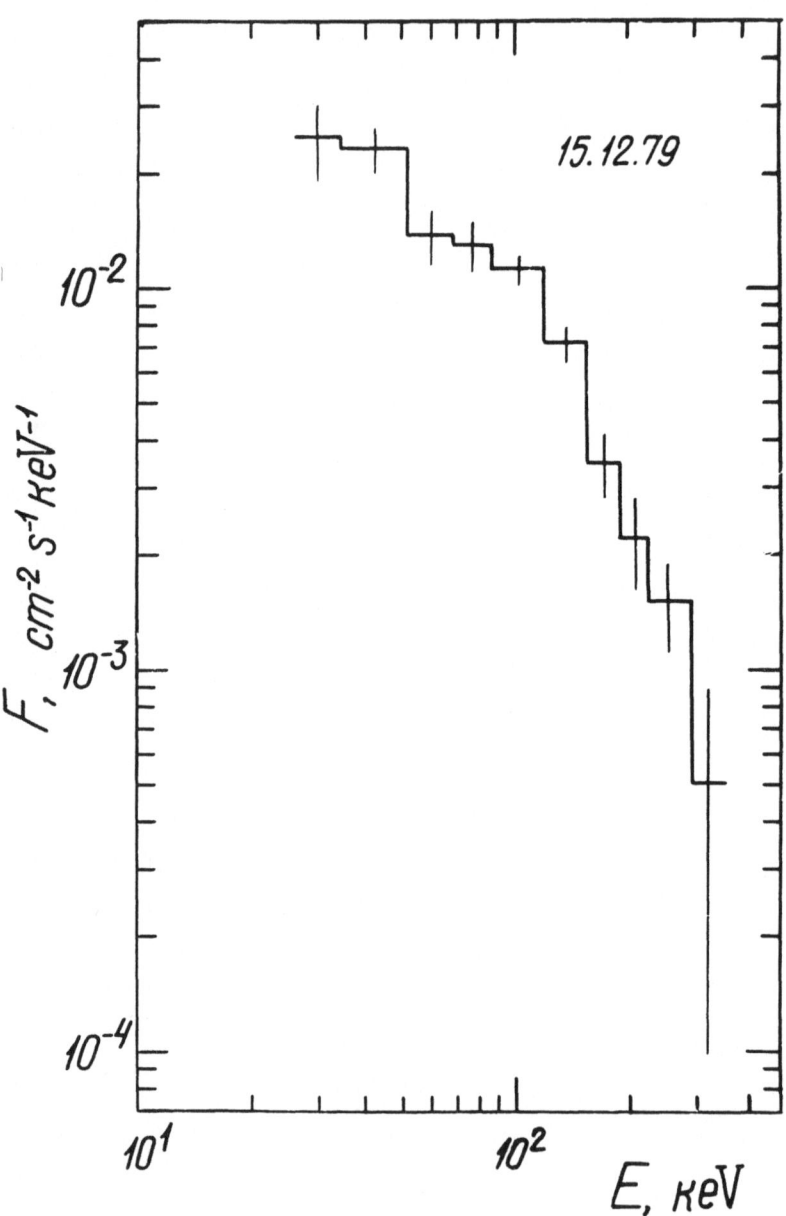

Fig. 28

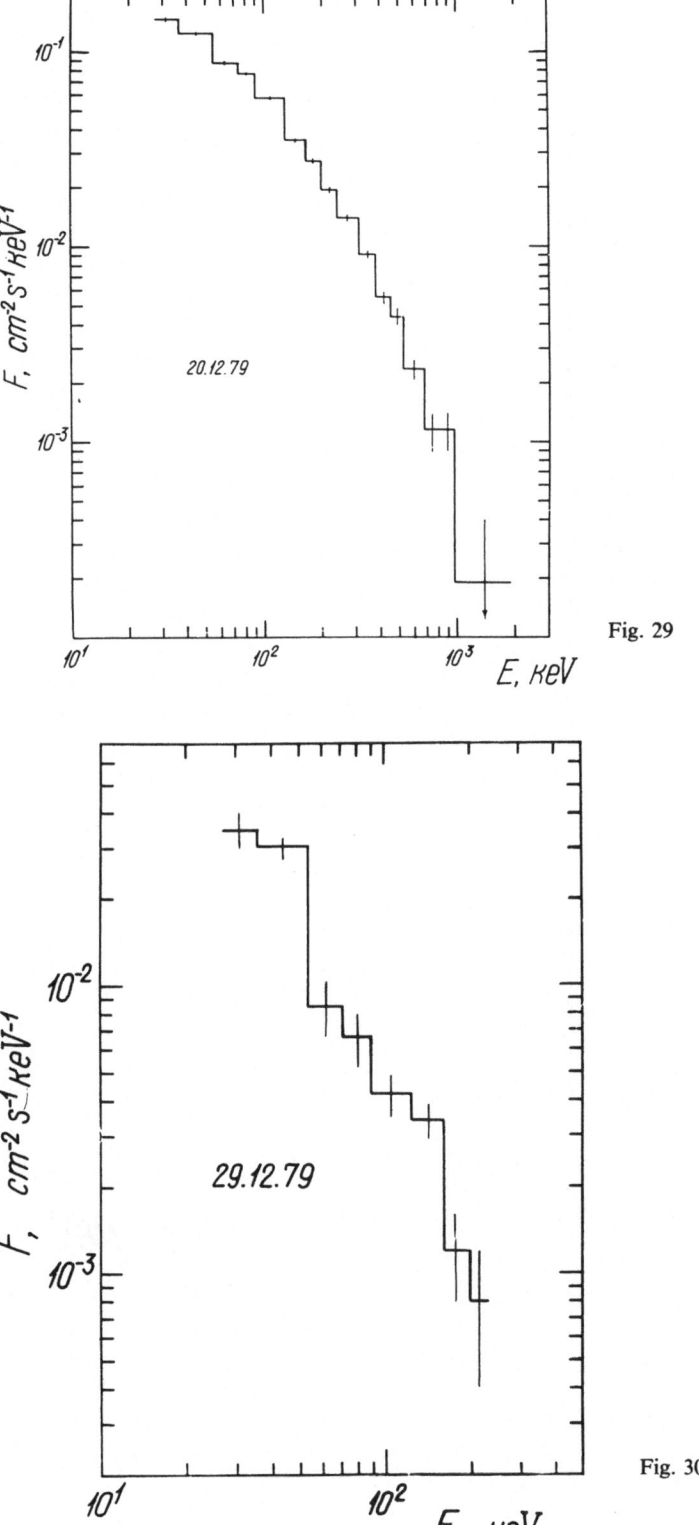

Fig. 29

Fig. 30

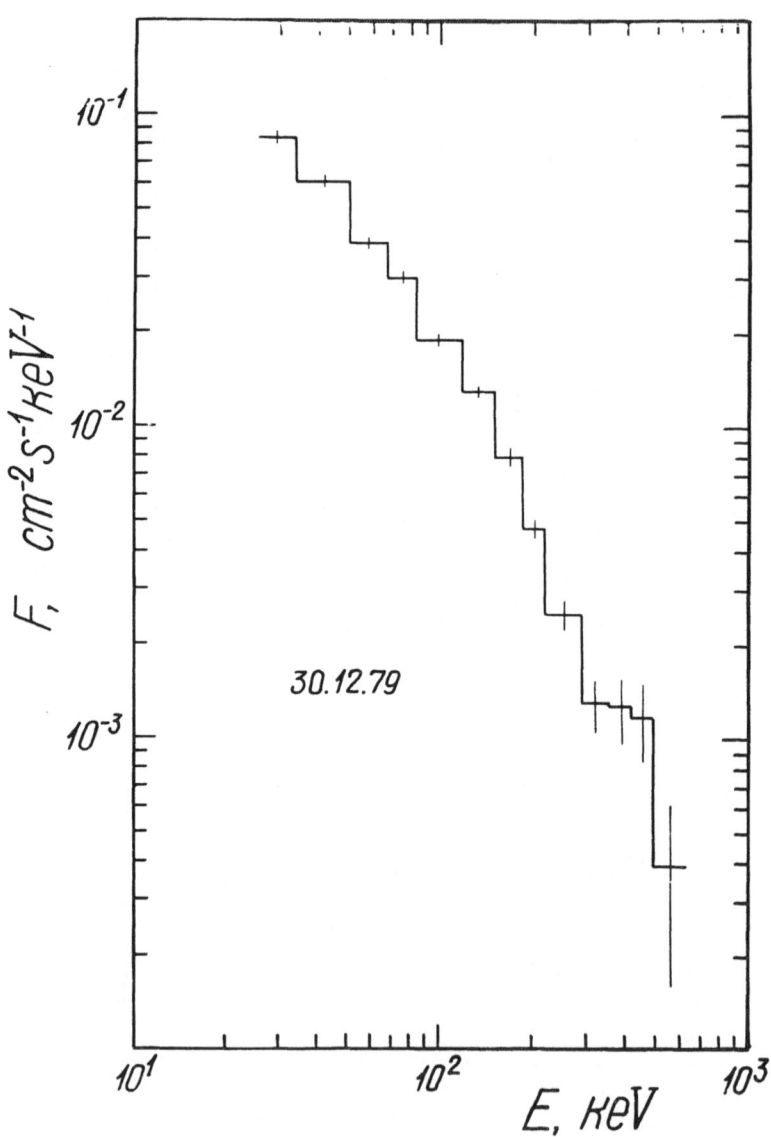

Fig. 31

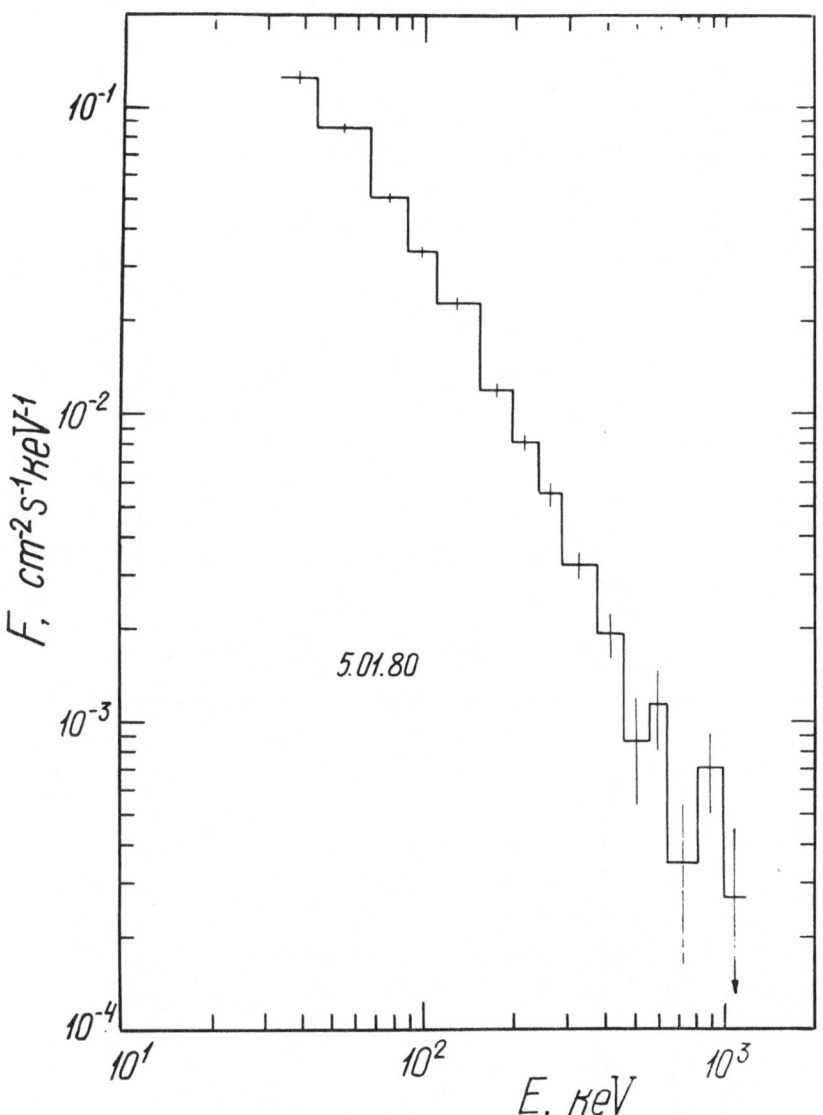

Fig. 32

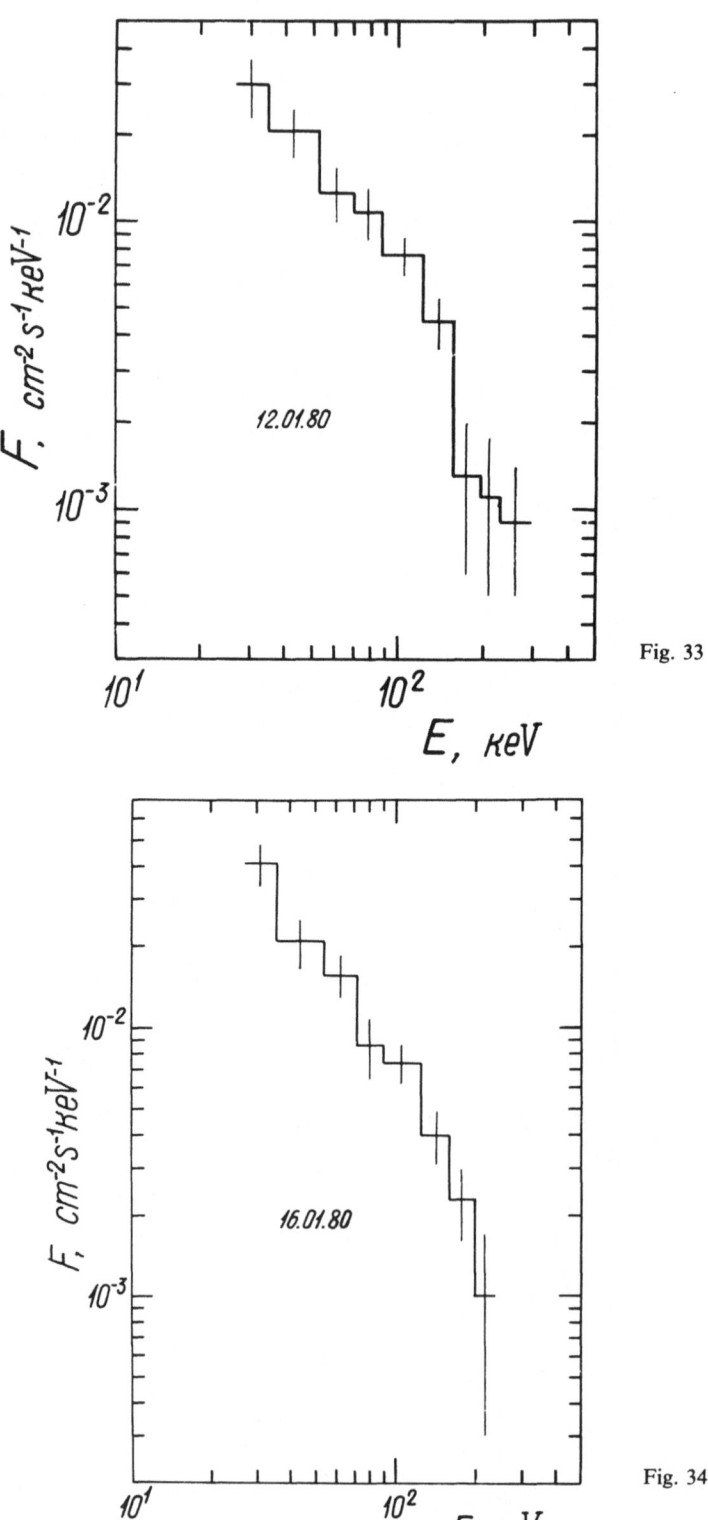

Fig. 33

Fig. 34

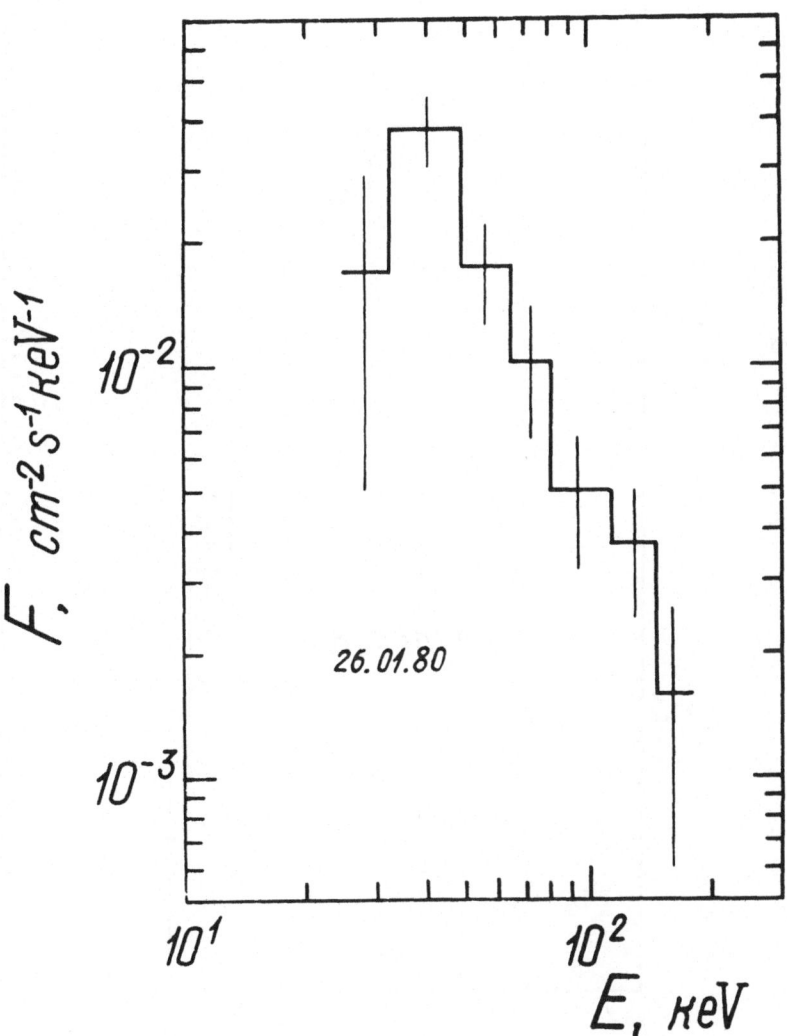

Fig. 35

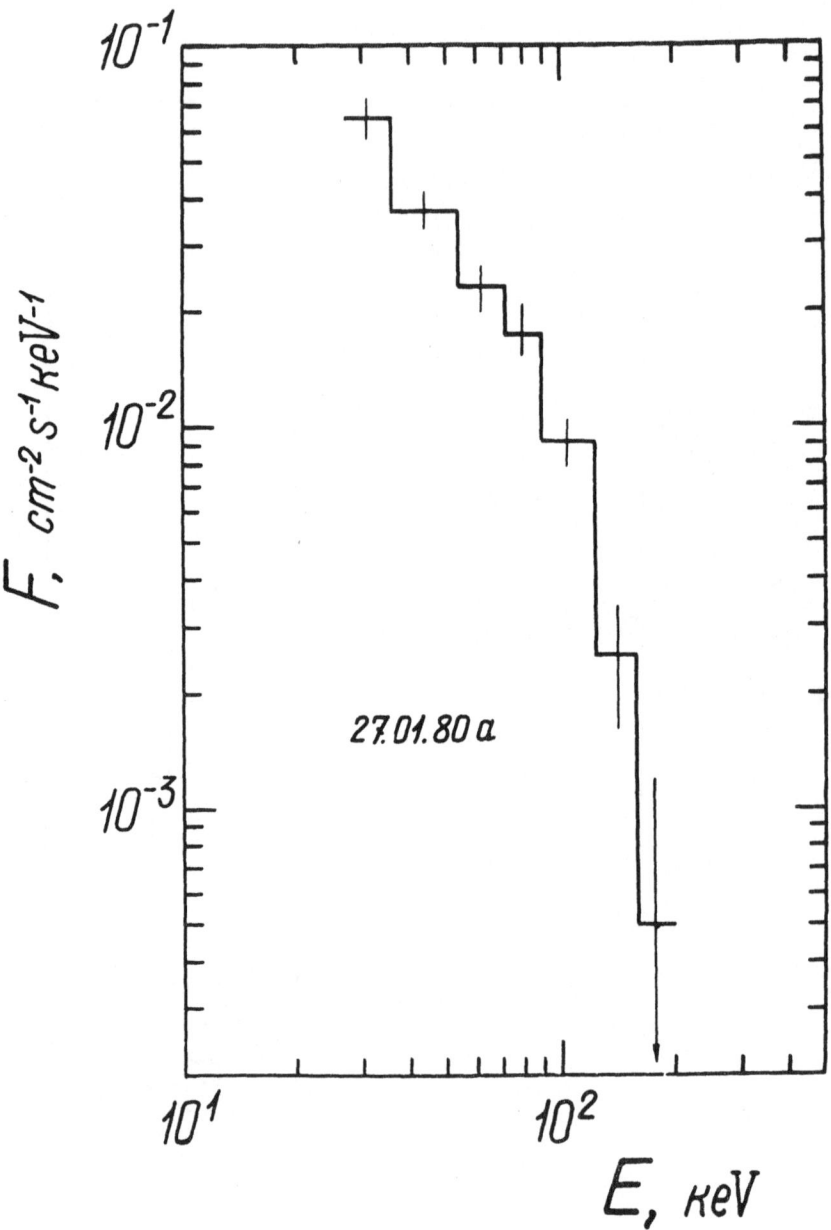

Fig. 36

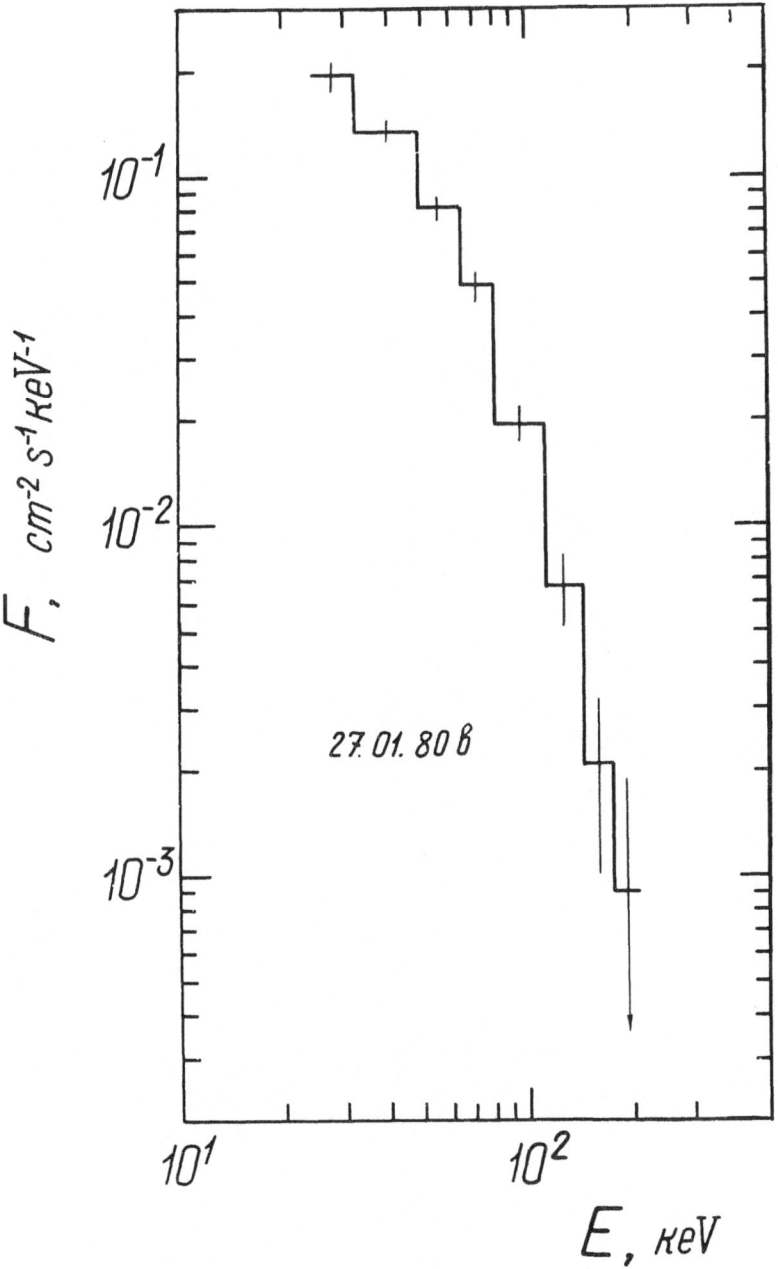

Fig. 37